Anti-apoptotic activity of the Herpes Simplex Virus Type 2 gene ICP10 PK.

Implications for the therapy of neurological disorders that involve apoptosis

by

Dana Perkins

Dissertation submitted to the faculty of the Graduate School
of the University of Maryland in partial fulfillment
of the requirements for the degree of
Doctor of Philosophy
2002

Dana Perkins
PhD Candidate, 2002

Professional Publications:

Perkins, D., Pereira, E.F.R., Gober, M., Yarowsky, P., and Aurelian, L. 2002. The Herpes Simplex Virus Type 2 gene R1 PK (ICP10 PK) blocks apoptosis in hippocampal neurons involving activation of the MEK/MAPK survival pathway. *J.Virol.* 76: 1435-1449

Perkins, D., Yu, Y.X., Bambrick, L., Yarowsky, P.J. and Aurelian, L. 2002. Expression of the Herpes Simplex Virus Type 2 Protein ICP10 PK protects neurons from apoptosis due to serum deprivation or genetic defects. *Exp. Neurol.* 174: 118-122

Perkins, D., Pereira, E.F.R. and Aurelian, L. 2001. A Herpes Simplex Virus Type 2 protein, ICP10 PK, inhibits caspase-3 activation in hippocampal neurons. *Promega's Cell Notes*, 2: 7-8

Perkins, D., Xing, Y.X., Bambrick, L.L., Yarowsky, P.J. and Aurelian, L. 2001. Herpes Simplex Virus Type 2 protein ICP10 PK promotes survival of trisomy 16 mouse hippocampal neurons. Abstract # 19.9 (oral presentation), Society for Neuroscience, 31st Annual Meeting, San Diego, CA, November 10-15.

Perkins, D., Smith, C.C. and Aurelian, L. 2001. Anti-apoptotic activity of HSV-2 protein ICP10 PK: novel strategy for treatment of Alzheimer's Disease. BioForum Technology Showcase, Maryland Bioscience Forum, October 3-4, Washington, DC.

Perkins, D., Pereira, E.F.R., Bambrick, L.L., Yarowsky, P. and Aurelian, L. 2001. Anti-apoptotic activity of HSV-2 protein ICP10 PK. Implications for therapy of neurodegenerative disorders. Abstract # 9.08 (platform presentation), 26th International Herpesvirus Workshop, July 28-Aug 3, Regensburg, Germany

Perkins, D., Pereira, E.F.R., Gyure, K., and Aurelian, L. 2001. The serotype-specific modulation of apoptosis determines Herpes Simplex Virus pathogenesis in the central nervous system. Abstract (platform presentation # 1), 20th Summer Symposium in Molecular Biology "Emerging Viral Diseases", June 13-16, Penn State University, University Park, PA.

Perkins, D., Pereira, E.F.R., Smith, C.C., and Aurelian, L. 2001. Herpes Simplex Virus Type 2 but not Type 1 has anti-apoptotic activity in hippocampal neurons: role of ICP10 PK. Abstract (oral presentation), 23rd Annual Graduate Research Conference, April 18, University of Maryland, Baltimore, MD.

Perkins, D., Pereira, E.F.R., Yarowsky, P.J., Xing, Y.-Y., and Aurelian, L. 2001. Anti-apoptotic activity of Herpes Simplex Virus Type 2 protein ICP10 PK. Implications for therapy of age-associated neurodegenerative disorders. Abstract # 5 (poster), University of Maryland Center for Research on Aging, 2nd Annual Aging Research Poster Day, March 1, University of Maryland, Baltimore, MD.

Perkins, D., Pereira, E.F.R., Smith, C.C., and Aurelian, L. 2000. ICP10 PK anti-apoptotic activity and CNS resistance to Herpes Simplex Virus Type 2 (HSV-2) infection. Abstract # 398.9 (oral presentation), Society for Neuroscience, 30th Annual Meeting, Nov. 4-9, New Orleans, La

Awards:

Winner of the 2nd Annual Research Paper Contest for Graduate Students, organized by the American Society for Microbiology, Maryland Branch, 2002.

Winner of the National Collegiate Inventors Competition (organized by the U.S. Patent and Trademark Office and the National Inventors Hall of Fame), 2001. Title of invention: "Gene therapy of Alzheimer's disease using a herpes simplex virus vector".

Patent co-author: "Anti-apoptotic activity of HSV-2 gene ICP10 PK". International Patent Application filed on July, 2001.

Travel grant to the 31st Annual Meeting of the Society for Neuroscience, November 10-15, 2001, San Diego, CA. Awarded by the Society for Neuroscience Chapters and Ely Lilly Company.

Travel grant to the 26th International Herpesvirus Workshop, July 28-August 3, 2001, Regensburg, Germany.

Travel grant to the Summer School/Workshop on "Brain Aging: Theories and Therapeutic Approaches", Elba International Neuroscience Program, September 14-26, 1991, Marina di Campo, Italy; organized by the Institute of Developmental Neuroscience & Aging, Denver, CO, and sponsored by the Institute for Research on Senescence, Sigma Tau, Pomezia, Italy.

Travel grant to the 14th Annual Meeting of the European Neuroscience Association, September 8-12, 1991, Cambridge, U.K.

ABSTRACT

Anti-apoptotic activity of Herpes Simplex Virus Type 2 protein ICP10 PK.

Implications for the therapy of neurological disorders that involve apoptosis

Dana Perkins, Ph.D., 2002

Dissertation Directed by Laure Aurelian, Ph.D., Professor

Department of Pharmacology and Experimental Therapeutics, School of Medicine,

University of Maryland

Apoptosis is an etiologic component of neurodegenerative disorders [e.g. Alzheimer's disease (AD), amyotrophic lateral sclerosis (ALS)] and acute brain injury (ischemia/hypoxia and trauma). Previous studies have shown that a Herpes Simplex Virus Type 2 (HSV-2) gene, ICP10 PK, activates the Ras/MEK/ERK pathway in non-neuronal cells. Because this pathway was implicated in neuronal cell survival, the present studies tested whether ICP10 PK blocks apoptosis in various experimental paradigms of neuronal apoptosis: 1) virus infection of primary hippocampal cultures, 2) trophic factor deprivation of NGF-dependent PC12 cells and primary hippocampal cultures, 3) naturally occurring genetic disorders such as the mouse trisomy 16 (Ts16), and 4) oxidative stress of N2a neuronal cells that express a mutant superoxide dismutase-1 (G85R). In the virus infection paradigm, HSV-1 and an HSV-2 mutant deleted in ICP10 PK (ICP10ΔPK) induced apoptosis, but apoptosis was not seen for HSV-2 and an HSV-2 mutant that retains the ICP10 PK (ICP10ΔRR), suggesting that ICP10 PK has anti-apoptotic activity. This activity was dependent on activation of the Raf/MEK/ERK survival pathway and inhibition of the pro-apoptotic JNK/c-Jun pathway. It involved inhibition of caspase-3 activation and PARP cleavage, likely resulting from induction of the anti-apoptotic protein Bag-1 and inhibition of the pro-apoptotic protein Bad. Ectopically expressed ICP10 PK inhibited apoptosis in the three other tested paradigms. The broad anti-apoptotic activity of ICP10 PK suggests that it may be used in gene therapy of neurological disorders that involve apoptosis. A replication-deficient HSV-2 mutant (ICP10ΔRR) was used for non-invasive delivery of ICP10 PK to the CNS. Expression of the therapeutic ICP10 PK gene following intranasal administration in the mouse, was consistent with a central spread of the vector through the central olfactory pathways to the hippocampus and related limbic structures. Collectively, the data suggest that ICP10 PK has broad anti-apoptotic activity and can be delivered to the CNS by peripheral administration of a replication-deficient HSV-2 vector.

ACKNOWLEDGMENTS

I am most grateful to my advisor, Dr. Laure Aurelian, and to Dr. Cynthia C. Smith who promoted and took part in this research, I am forever indebted for their invaluable suggestions and generosity. Without them, the anti-apoptotic activity of ICP10 PK and its potential in therapy would have never come to the public attention.

To Drs. Edna F.R. Pereira, Linda Bambrick, Kymberly Gyure and Paul J. Yarowsky, I wish to express my deepest gratitude for their contribution to these studies.

To Paul H. Lackey, who helped preparing my presentations at various scientific meetings, my warmest thanks.

TABLE OF CONTENTS

VIRUS NOMENCLATURE

HSV-2   Herpes Simplex Virus Type 2, wild type.

HSV-2 (R)   Herpes Simplex Virus Type 2, revertant.

ICP10ΔPK   HSV-2 mutant that lacks the PK domain of the ICP10 gene.

ICP10ΔRR   HSV-2 mutant that has the RR domain of the ICP10 gene replaced with
                the *E. coli. LacZ* gene.

HSV-1   Herpes Simplex Virus Type 1, wild type.

ICP6Δ   HSV-1 mutant that lacks 90 % of the coding sequence of the ICP6 gene.

*hr*R3   HSV-1 mutant that retains only 38 % of the N-terminus domain of ICP6
                gene and has the RR domain of the ICP6 gene replaced with the *E. coli.*
                LacZ gene.

I   INTRODUCTION

The present studies will examine the role of a Herpes Simplex Virus Type 2 gene, ICP10 PK, in apoptosis. The purpose of the Introduction Chapter is to review the apoptotic pathways and their relationship to the major cellular signaling pathways, as well as the specific features of Herpes Simplex Viruses as they relate to apoptosis.

A.   Ras and its major effectors

Signal transduction pathways relay the information provided by growth factors or other extracellular stimuli to the nucleus where the response takes the form of specific gene expression. Signals can be interpreted as mitogenic, differentiating or apoptotic, according to the specific response they elicit. They act by inducing a cascade of phosphorylation events that results in altered protein conformation and/or enzymatic activity. Growth factors (or other ligands) bind to cell surface receptors (TKR, receptor tyrosine kinase) and activate their intrinsic tyrosine kinase activity. In one of the major signaling cascades (Fig. 1), activated receptor binds the adaptor protein Grb 2 and the guanine nucleotide exchange factor Sos (Aronheim *et al.*, 1994).
Tyrosine phosphorylation of the adaptor protein Shc followed by association with the Grb2-Sos complex also occurs (Sasaoka *et al.*, 1994a; Sasaoka *et al.*, 1994b). Binding of Sos to adaptor protein(s) brings this protein in the vicinity of Ras which is membrane-

bound. The guanine nucleotide exchange factor catalyzes the exchange of GDP to GTP, thereby the formation of the active form of Ras (Ras-GTP). The negative regulator of Ras [Ras GTP-ase activating protein (Ras-GAP, also known as p120 GAP)] accelerates the intrinsic GTP-ase activity of Ras thereby promoting the formation of the inactive GDP bound Ras. Ras has multiple effector proteins which form distinct signaling pathways (reviewed by Vojtek et al., 1998). The common denominator for this multitude of effector proteins is their affinity for the active GTP-bound Ras. The mitogen-activated protein kinases (MAPKs) and phosphoinositide 3-kinase (PI3-K) pathways, are activated by a wide range of stimuli that have an impact on proliferation, differentiation and survival. MAPKs comprise 3 families of serine-threonine protein kinases: p44/p42 extracellular signal-related kinases [MAPK1/2, also known as extracellular signal regulated kinases (ERK1/2)], c-jun N-terminal protein kinase (JNK)/stress activated protein kinase (SAPK) and p38 MAP kinase (Vojtek et al., 1998). The MAPK (ERK) pathway is initiated when Ras-GTP binds and activates the serine-threonine (Ser-Thr) kinase Raf. The family of Raf-protein kinases consists of A-Raf, B-Raf and c-Raf isoforms that have overlapping and unique regulatory functions. c-Raf-1 is ubiquitously expressed, with highest expression levels in striated muscle, cerebellum and fetal brain, whereas A-Raf and B-Raf are differentially expressed with highest levels in neuronal and urogenital tissues (Hagemann and Rapp, 1999). MEK is so far the only in vivo substrate common for all Raf proteins. Originally it was found to become activated by c-Raf-1 (Blenis, 1993). However, in bovine brain extracts, NIH 3T3 cells and NGF-dependent PC12 cells, B-Raf is the main MEK activator (Hagemann and Rapp, 1999). Raf phosphorylates/activates the dual specificity kinase, MEK, which phosphorylates ERKs on both threonine (Thr) and tyrosine (Tyr) residues. Activated ERKs induce gene expression by activation of a ternary complex transcription factor (TCF), such as Elk-1 or serum response factor accessory protein (SAP), that forms a complex with a dimer of serum response factor molecules (SRF) to regulate gene expression through the serum response element (SRE). ERKs may also regulate gene expression by activating the c-myc/N-myc transcription factors. Moreover, ERKs may phosphorylate/activate the pp90 ribosomal S6 kinases (RSK). Complete RSK activation requires additional phosphorylation events mediated by PDK1 (phosphoinositide-dependent kinase 1) suggesting an integration of regulatory inputs from both the ERK and PI3-K/PDK1-dependent signaling pathways. RSKs (also referred to as MAPKAP kinases-1) family consists of 3 isoforms (RSK1,2,3) that are widely expressed in higher eukaryotes and are all specifically activated through phosphorylation by ERK but not by any other MAPK subfamilies. The three members of the RSK family are more than 80 % identical in amino acid sequence, and translocate to the nucleus after their activation (Xing *et al.*, 1996). Activated RSKs activate in turn transcription factors such as CREB (cAMP-response-element-binding protein), SRF or c-fos (Vojtek *et al.*, 1998; Bonni *et al.*, 1999) and phosphorylate the transcriptional inhibitor IκB, thereby leading to activation of NFκB transcription factor (Schouten *et al.*, 1997). Thus, a bifurcation exists in the Ras/Raf/ERK pathway, both branches of which contribute to specific gene transcription. One of these induced genes, c-fos, is a component of the sequence specific transcriptional activator AP-1 (Angel and Karin, 1991). AP-1 is a collection of dimers composed of jun, fos or ATF families of bZIP (basic region-leucine zipper) DNA binding proteins. These dimers bind to a common cis acting element known as the TRE (TPA response element) or the AP-1 site. While many of the AP-1 factors are transcriptional activators, certain AP-1 complexes can function as transcriptional repressors. The exact function and potency of AP-1 complexes is determined by their composition, interaction with other transcription factors (such as Ets or Rel) or coactivator proteins (such as JAB1) (Minden and Karin, 1997).

The PI3-K pathway begins with the binding of Ras-GTP to the catalytic subunit of PI3-K leading to accumulation of the second messenger phosphatidylinositol 3,4,5-triphosphate (PIP3). One of the downstream effector of PI3-K, namely Akt (also known as protein kinase B) is activated by binding to PIP3 which leads to its localization to the plasma membrane (Hemmings, 1997). However, its enzymatic activity is further increased by subsequent phosphorylation events (Stephens *et al.*, 1998). Akt has three cellular isoforms, of which c-Akt3/Rac-PKγ is the major species expressed in neurons (Datta *et al.*, 1999). The kinases (3-phosphoinositol-dependent kinases, PDK1,2) are themselves regulated by PIP3, suggesting that the lipid products generated by the PI3-K enzymes control the activity of Akt by regulating its localization and activation. Akt has been shown to affect, directly or indirectly, three transcription factors: Forkhead 1 (FKHRL1), CREB and NFκB. FKHRL1 is a transcription factor that induces apoptosis by increasing the levels of the Fas ligand and is phosphorylated/inactivated by Akt (Brunet *et al.*, 1999). The phosphorylation of CREB and IκB kinase (IKK, phosphorylates IκB thereby activates NFκB) is associated with specific gene transcription and cell survival (Du and Montminy, 1998; Brunet *et al.*, 1999; Kane *et al.*, 1999; Riccio *et al.*, 1999).

The JNK/SAPK and p38 MAPK stress activated pathways start with activation of

MEK kinase (MEKK) by Ras or Rac. In turn, it phosphorylates the MEK homolog (MKK, also known as SEK) which phosphorylates/activates JNK and p38 MAPK. There are 10 identified isoforms of JNK originating from 3 homologous genes (JNK1, JNK2 and JNK3, also known as SAPKα, SAPKβ and SAPKγ) with molecular masses of 46, 54 and 49 kDa, respectively, due to alternative splicing (Kyriakis *et al.*, 1995; Gupta *et al.*, 1996). In contrast to MAPK (ERK) pathway, Ras activation alone stimulates only a low level of JNK activity, implying distinct differences between the upstream JNK and ERK pathways (Jessel and Goodman, 1996). Once phosphorylated/activated, JNK activates a number of transcription factors, such as c-Jun (by phosphorylation on serines 63 and 73) whose function has been implicated in various cellular events ranging from cell proliferation and differentiation to neoplastic transformation (Angel and Karin, 1991). In addition to c-Jun, the JNKs also activate other transcription factors such as ATF-2, TCF/Elk-1, c-fos and p53 (Minden and Karin, 1997). Interestingly, c-Jun and ATF-2 dimerize and bind to the c-Jun promoter, thereby stimulating the expression of the c-Jun gene. JNK also mediates c-fos induction by phosphorylation of Elk-1 (which is one of the best characterized TCFs) in the same manner as ERKs. While induction of the c-fos gene by growth factors is mediated primarily by the ERKs, the JNKs are most likely responsible for c-fos induction in response to cellular stress and cytokines. In both cases, induction of the c-fos gene leads to increased expression of the c-fos protein, which in turn can translocate to the nucleus and form heterodimers with c-Jun which are more stable than Jun-Jun homodimers leading to a more stable AP-1 complex (Smeal *et al.*, 1989). P38 MAPK is activated by many of the extracellular stimuli that activate JNK as well. The upstream activators pf p38 MAPK include MKK3 and MKK6 but also ASK1 and JNKK (Minden and Karin, 1997). Like JNK, p38 can phosphorylate and activate ATF-2 (Raingeaud *et al.*, 1996) thereby contributing to the c-Jun gene induction mediated by the AP-1 binding site in the c-Jun promoter. In addition, p38 phosphorylates and activates MEF2C transcription factor which binds to the MEF2 site in the c-Jun promoter (Fig.2). Also, p38 can phosphorylate and activate TCF-Elk-1 which mediates c-fos induction by binding together with SRF to the SRE. Under some circumstances, c-fos induction also requires another element (CRE), which is recognized by the transcription factor CREB (Bonni *et al.*, 1995). p38 was shown to lead to phosphorylation/activation of CREB by MAPKAP2, a p38 activated kinase (Tan *et al.*, 1996).

There is considerable crosstalk and cooperation between signaling pathways whether they originate or not from Ras. Thus activation of ERK and JNK signaling pathways is Ras-dependent in response to growth factors and Ras-independent in response to other stimuli such as cytokines (Minden *et al.*, 1997) or estrogen (Improta-Bears *et al.*, 1999). However, it has been suggested that the pathways cooperate in inducing oncogenic transformation (Plattner *et al.*, 1999). Also, the ERK and PI3-K cascades may cooperate or function independently in promoting cell survival. They are particularly important in neuronal survival since both MEK/ERK and PI3-K/Akt convey the neurotrophin signals. Whereas the PI3-K/Akt pathway always appears to have a protective effect, the MEK/ERK pathway is not always anti-apoptotic. Under some circumstances (depending on the cell type and phase of cell cycle) activation of MEK/ERK induces apoptosis. For example, ERK activation in Swiss 3T3 fibroblasts results in apoptosis in S phase but the arrest of cells in other parts of cell cycle (Fukasawa *et al.*, 1995). By contrast, the activation of JNK/SAPK and p38 MAP kinases by apoptotic signals that leads to increased c-Jun and p53 synthesis as well as c-Jun phosphorylation/activation induces cell death or apoptosis in most cell types (Ham *et al.*, 1995; Kyriakis and Avruch, 1996 ; Morishima *et al.*, 2001).

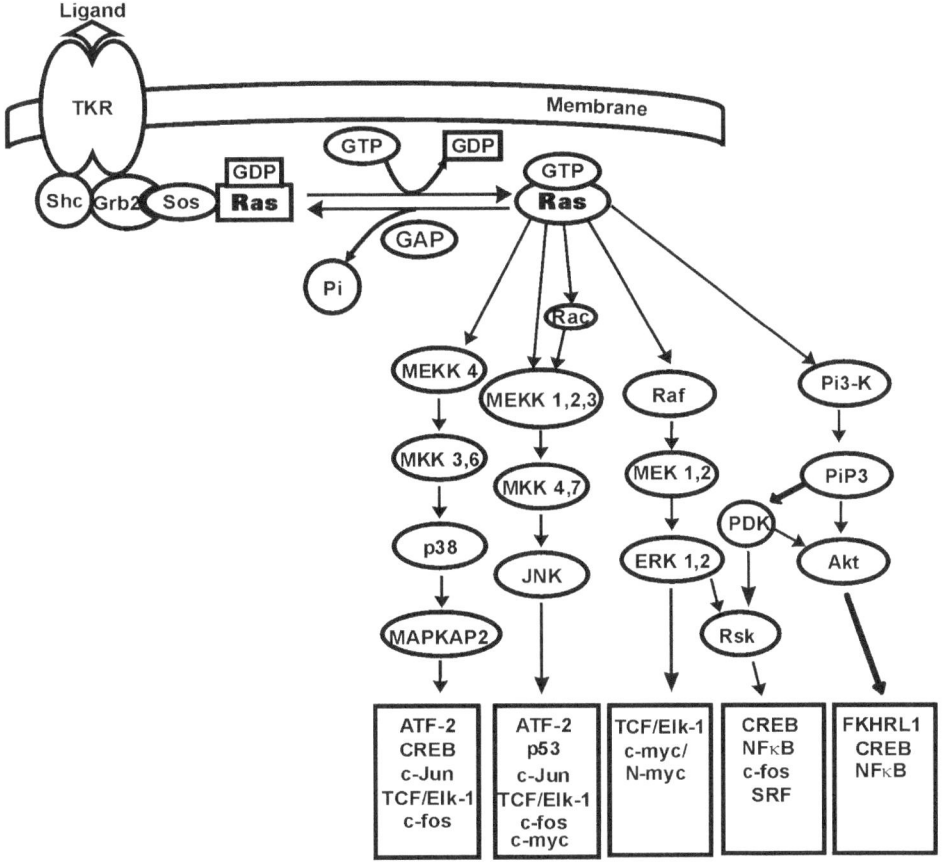

Fig. 1. **Major signaling pathways activated by Ras.** Upon ligand binding to a tyrosine kinase receptor (TKR), adaptor proteins (Shc, Grb2) are recruited to the receptor and form a complex with Sos (a nucleotide exchange factor). Sos activates Ras by catalyzing the exchange of GDP to GTP. GAP negatively regulates Ras by catalyzing the hydrolysis of GTP to GDP. Activated Ras (Ras-GTP) activates signaling pathways (Raf/MEK1,2/ERK; PI3-K/Akt; MEKK4/MKK3,6/p38; MEKK1,2,3/MKK4,7/JNK). Signaling pathway activate in turn transcription factors (such as CREB, c-fos, c-Jun, ATF-2, TCF/Elk-1, etc.) leading to specific gene transcription.

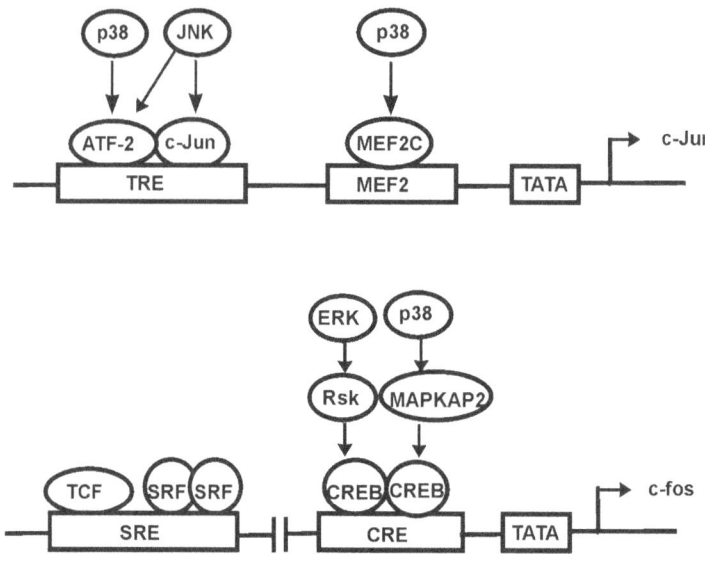

Fig. 2. **Regulation of c-fos and c-Jun genes by MAPKs.** ERK, JNK and p38 MAPK converge on c-Jun and c-fos genes by activating transcription factors that bind to cis-acting elements in their promoters.

1.   **The role of Ras/Raf/MEK/ERK pathway in the nervous system**

The ERK cascade has been classically studied as a critical signaling pathway involved in neoplastic transformation events. Although this cascade is typically studied in the context of mitotic cell regulation, its components are actually most abundant expressed in postmitotic neurons of the developed nervous system (Boulton *et al.*, 1991; Fiore *et al.*, 1993). Many signals converge on ERKs. Growth factors activate ERKs in a Ras-dependent manner, whereas protein kinase C (PKC) and c-mos kinase activate ERKs in a Ras-independent manner (Iida *et al.*, 2001). Several studies that used cultured neuronal cell lines described calcium- and c-AMP-induced activation of ERKs either as Ras-dependent (Rosen *et al.*, 1994), or Ras-independent [mediated by Rap1, a Ras-related small GTP-ase (Grewal *et al.*, 2000)]. It was also reported that NGF induces the rapid and transient activation of ERKs in a Ras-dependent manner and the prolonged activation of ERKs through Rap1 activation (York *et al.*, 1998). Studies that used dominant negative forms of Ras or Gap1 suggested that in hippocampal neurons, ERK activation is Ras-dependent and Rap1-independent in response to NMDA, calcium ionophore, membrane depolarization, forskolin and BDNF (Iida *et al.*, 2001). By contrast, CREB phosphorylation/activation depended on multiple pathways, as evidenced by its inhibition by dominant negative Ras in BDNF- but not NMDA-signaling (Iida *et al.*, 2001).

ERKs regulate a diverse array of functions in neurons including neurotrophic, differentiation, and survival actions of neuronal growth factors, synaptic plasticity and long-term potentiation.

1.1.   ERKs as effectors of neurotrophin actions

During development, many of the same peptide growth factors that regulate the proliferation of non-neuronal cells also stimulate the proliferation of neuroblasts. Insulin-like growth factors (LeRoith *et al.*, 1993), fibroblast growth factors (Unsicker *et al.*, 1993), the epidermal growth factors, and the neurotrophins (Morrison, 1993) have all been shown to enhance proliferation of neuronal precursor cells. A number of these growth factors can promote the survival and differentiation of distinct, but overlapping subsets of neurons. In all cases these growth factors function by binding and activating specific receptor tyrosine kinases. This then triggers the activation of cellular signaling cascades (such as ERKs) that culminate in specific programs of gene transcription and particular cellular responses.

Neurotrophins trigger a variety of biological responses, including proliferation, differentiation, and survival of neuroblasts, and survival and adaptive responses of mature neurons. Among the differentiation responses that neurotrophins elicit are enhanced neurite outgrowth (Koo and Liebl, 1992; Cohen *et al.*, 1994), changes in electrophysiological properties of neurons (Levine *et al.*, 1995), and alterations in neuronal cell fate (Sieber-Blum, 1991).

The specific receptor tyrosine kinases bound and activated by neurotrophins belong to the Trk family of receptors. TrkA, TrkB, and TrkC are respectively the receptors for the nerve growth factor (NGF), brain-derived neurotrophic factor (BDNF), and neurotrophin 3 (NT3) (Segal and Greenberg, 1996). The TrkB receptor is also bound and activated by neurotrophin 4/5 (NT4/5) (Bothwell, 1991). In addition to binding to a specific Trk, each of the neurotrophins mentioned above interacts with a common receptor, p75, which is structurally related to cytokine receptors (Rodriguez-Tebar *et al.* 1992). NGF binding to the p75 receptor, in the absence of Trks, was found to activate the sphingomyelinase pathway (Dobrowsky *et al.* 1994) and also to generate specific biological responses, such as change in cell-to-cell adhesion (Itoh *et al.*, 1995).

The major signaling pathways activated by Trk receptors are Ras/Raf/ERK and PI3-K/Akt. PI3-K enzymes are normally present in the cytosol and can be activated directly by recruitment to an activated Trk receptor, or indirectly through activated Ras. Active PI3-K enzymes catalyse the formation of the lipid 3'-phosphorylated phosphoinositides (PIP3), which regulate the localization and activity of a key component in cell survival, the Ser-Thr kinase Akt (Philpott *et al.*, 1997). Neurotrophins also stimulate the docking of the adaptor protein Shc to activated Trk receptors, which in turn triggers the activation of Ras and the downstream Raf/MEK/ERK cascade (Bonni *et al.*, 1999). Ras activation is a critical event in neurotrophin induced differentiation. When microinjected in rat pheochromocytoma (PC12) cells, a mutated constitutively active form of H-Ras induces PC12 cell differentiation into a sympathetic, neuron-like phenotype and thus mimics the action of NGF (Bar-Sagi and Feramisco, 1985). This effect is associated with sustained activation of ERKs and a role for a Ras-dependent pathway has been well established. In PC12 cells, the downstream effector of activated Ras and thus of NGF signal is Raf-B (Oshima *et al.* 1993). Once activated, Raf-B phosphorylates and activates MEK1. Expression of a constitutive form of MEK1 was sufficient to induce neurite outgrowth in PC12 cells, whereas expression of a dominant negative form of MEK1 blocked NGF-induced outgrowth (Cowley *et al.*, 1994). The only known targets of MEK1 are ERK1,2. Phosphorylation/activation of ERKs by MEK1 leads to specific gene transcription either by direct activation of specific transcription factors or indirectly, by ERK activation of RSK2 which in turn activates specific transcription factors.

Neurotrophins, by virtue of the signaling pathways that they activate, are also play a central role in controlling the neuron number during development. The decision whether an individual neuron should live or die during the development of the nervous system is determined primarily by the availability of neurotrophic factors, although other influences (e.g. afferent input) are also important. For example, NGF can prevent neuronal death occurring during development and after target removal or axotomy (separation of the neuronal cell body from target). Similarly, administration of NGF neutralizing antibodies to immature animals results in massive death of specific neuronal populations, suggesting a role for NGF in neuronal survival (Levi-Montalcini and Booker, 1960).

A central role for the PI3-K/Akt in neuronal survival was first suggested by the observation that PI3-K inhibitors block the survival effect of NGF (Yao and Cooper, 1995). Also, the downstream target of PI3-K, Akt, was shown to have an essential role in neuronal survival, as demonstrated by the finding that active Akt supports neuron survival in the absence of trophic factors, whereas a dominant negative mutant of Akt inhibits neuronal survival even in the presence of survival factors (Datta *et al.*, 1999).

The studies showing that overexpression of MEK supports PC12 cell survival in the absence of NGF indicated that the MEK/ERK pathway is also important for neuronal survival (Xia *et al.*, 1995). MEK/ERK activity was also required for survival promoted by BDNF or insulin in cerebellar neurons (Bonni *et al.*, 1999). Recently, more consideration was given to the role of the Raf/MEK/ERK pathway in protecting neurons from death due to injury, stress or toxicity, rather than protecting from growth factor withdrawal (Kaplan and Miller, 2000). For example, the glial cell line-derived neurotrophic factor (GDNF), which was first characterized for its trophic activity on dopaminergic neurons, exerts a neuroprotective effect on cortical neurons via activation of ERK pathway leading to a reduction of the NMDA (a ion-channel-linked glutamate receptor)-mediated Ca^{2+} influx (Olivier *et al.*, 2001).

Apart from regulation of developmental survival and growth of neurons, neurotrophins also seem to play a role in the maintenance of function of mature neurons. For example, NGF has been shown to be a potent survival factor for cholinergic basal forebrain neurons (Hefti *et al.*, 1989) which consistently degenerate in Alzheimer's disease (AD) (Mufson *et al.*, 1989). BDNF may also play a role in the maintenance of these neurons as well as dopaminergic neurons of the substantia nigra, which degenerate in Parkinson's disease (PD). Also, GDNF is a survival factor for substantia nigra dopaminergic and locus coeruleus noradrenergic neurons, which are affected in both PD and AD. This neurotrophin, as well as ciliary neurotrophic factor (CNTF) can affect those spinal motor neurons that degenerate in amyotrophic lateral sclerosis (ALS) (Lapchak *et al.*, 1996). Administration of neurotrophic factors can attenuate age-related and experimentally induced degeneration and behavioral deficits in animals, suggesting that such molecules may become useful in the treatment of neurodegenerative disorders (Hefti and Knusel, 1990). However, the side effects associated with administration of neurotrophic factors limited their effectiveness and lead to only modest results in human trials. For example, long term administration of NGF in humans may induce aberrant cholinergic sprouting, as observed in lesioned rodents (Williams *et al.*, 1986). The possibilities of undesired proliferation of NGF-responsive non-neuronal cells and of changes in local blood flow due to NGF action on intracerebral sympathetic neurons have to be considered. Furthermore, administration of NGF to adult brain may actually trigger pathological changes typical for AD since intraventricular NGF administration to neonatal animals has been shown to elevate brain levels of amyloid precursor protein mRNA (Mobley *et al.,* 1988). These obstacles in neurotrophic factor therapy of neurodegenerative disorders suggest the need for novel therapeutic approaches for neurodegenerative disorders which circumvent the side effects discussed above.

1.2. ERKs in synaptic plasticity and long term potentiation

Neuronal cells have a number of properties that change during or after stimulation. For example, sensory neurons adapt; their response tends to fade with continued or repeated stimulation. Similarly, repeated stimulation of motor neurons may cause a muscle to give either stronger (facilitation) or weaker (depression) responses. These plastic changes that occur at synapses as a consequence of activity may be found at many sites in the nervous system and these changes may also contribute to learning and memory- which are basic properties of central systems that control the behavior of the whole organism. Learning (the ability to change with experience) is intimately linked to memory (the capacity to store and recall those changes). The working memory mechanisms operate in short time periods (seconds) and are responsible for language understanding and short-term storage of trivial information that is used during the day and forgotten shortly after. The prefrontal cortex within the frontal lobe plays a critical role in working memory (Shepherd, 1994). By contrast, the long-term memory, or associative memory, is involved in building up the information that provides for basic knowledge, skills, and ways of thinking and behaving, and it may take minutes, years or a lifetime. The main model that has emerged in recent years for analyzing and understanding the mechanisms underlying long-term memory is long-term potentiation (LTP) in the mammalian hippocampus. The term LTP was coined by Bliss and Lomo in 1973, when they recorded the field potentials evoked in the dentate fascia of anesthetized rabbits by a shock to the entorhinal cortex. They stimulated tetanically (at high frequencies) for several seconds, then tested with a single shock at various intervals after that, and found that the part of the recording due to the synaptic response of the granule cells of the hippocampus grew to a much greater amplitude than normal, a phenomenon then called LTP. In other words, LTP is a form of synaptic plasticity defined as a long-lasting, use-dependent increase in synaptic efficacy, which can be elicited with brief pulses of high-frequency stimulation. In the dentate cells of the hippocampus, LTP lasts for hours, or even days and weeks. It has subsequently been demonstrated in hippocampal pyramidal cells and other types of neurons as well (Shepherd, 1994). LTP can be divided in two stages: early LTP (E-LTP), which lasts < 2 hrs and does not require *de novo* protein synthesis, and late LTP (L-LTP) which is sensitive to inhibitors of transcription and translation.

ERKs are excellent candidates for regulators of synaptic plasticity in post-mitotic neurons of the adult nervous system due to their presence in soma and dendrites of pyramidal neurons of the adult brain and their ability to modulate gene transcription and/or other targets. Compelling evidence for a role for ERKs in learning and memory has come from rodent behavioral studies, where inhibition of ERK specifically impaired learning and behavioral performance was associated with increased ERK activity (Atkins *et al.*, 1998; Berman *et al.*, 1998; Blum *et al.*, 1999). Hippocampal LTP-inducing stimuli specifically increase the expression of two components of the ERK cascade, ERK2 and B-Raf (Thomas *et al.*, 1994). Other studies suggested simultaneous activation of both ERK1 and ERK2 (Kurino *et al.*, 1995). Moreover, ERKs are required for LTP induction, as demonstrated by the fact that inhibition of the ERK cascade using pharmacological inhibitors markedly attenuates the induction of LTP in pyramidal cells (English and Sweatt, 1997). ERK activity is necessary even for E-LTP, suggesting a role for ERK in post-translational processes underlying LTP induction and maintenance. ERKs may participate in both E-LTP and L-LTP, first by modifying existing proteins that determine synaptic behavior, and subsequently by regulating the expression of the proteins necessary for the maintenance of synaptic changes. For example, ERKs induce both the post-translational modification and increased expression of a central protein in neuronal plasticity, Ca^{2+}/calmodulin-dependent protein kinase II (CaMKII), after the induction of

LTP (Giovannini *et al.*, 2001). CaMKII is an important component of the postsynaptic density, a complex of proteins that include many members of the synaptic signaling network (Husi *et al.*, 2000), including AMPA- and NMDA-type glutamate receptors (which are phosphorylated by CaMKII) (Gardoni *et al.*, 2001).

Induction of LTP in the CA1 region of the hippocampus usually requires calcium influx through postsynaptic NMDA receptor channels and enhancement of glutamate release presynaptically is thought to be one of the main mechanisms of LTP. A candidate for the retrograde messenger from the postsynaptic to the presynaptic neuron is believed to be nitric oxide (NO). As such, direct application of NO to CA1 neurons mimics LTP (Sasaki *et al.*, 2000). Evidence for the involvement of NO in NMDA receptor-signaling came from experiments that showed that stimulation of NMDA receptors in cultured neurons activates the Ras/ERK pathway leading to CREB phosphorylation via calcium dependent activation of NO-synthase (NOS) enzyme and NO production (Yun *et al.*, 1998).

Another important role of ERKs in learning and memory comes from their involvement in signal transduction from activated muscarinic (mAChR) and nicotinic (nAChR) acetycholine receptors, which are involved in learning and memory (Messer *et al.*, 1991). The cholinergic innervation of the cerebral cortex and the hippocampus originates primarily from the cholinergic basal nuclear complex. Lesions of these basal forebrain neurons result in impairment in memory, learning and attention. Muscarinic cholinergic agonists facilitate learning and memory (Jerusalinsky *et al.*, 1997). It has been suggested that improvement of cognitive functions by muscarinic agonists is the result of their modulation of ERK activity. ERK activation is induced by muscarinic agonists in the neocortex and hippocampus *in vivo*, in hippocampal slices, and in primary cortical cultures (Rosenblum *et al.*, 2000). nAChRs are widely distributed in the CNS (Clarke *et al.*, 1985). One class of nAChR, i.e. nAChRs, that bind and are functionally blocked by, the snake venom toxin α-bungarotoxin (α-BGTX) and contain α7 subunits (α7 nAChR) has received special attention. Hippocampal α7 nAChR are involved in neuronal plasticity and memory formation (Albuquerque *et al.*, 1997; Perry *et al.*, 1995) and are particularly susceptible to the ravages of AD. It has been shown that the nicotine-binding sites are reduced in the brains of patients with AD, PD and DS, and [^{125}I]α-BGTX-binding sites are decreased in selected brain regions with AD (Court *et al.*, 2001; Hilmas *et al.*, 2001). Moreover, it was reported that Aβ binds to α7 nAChR (Wang *et al.*, 2000), and this may inhibit α7 nAChR-dependent learning and memory. Significantly, recent studies showed that stimulation of α7 nAChR by nicotine leads to activation of ERK (Dineley *et al.*, 2001) and PI3-K/Akt (Kihara *et al.*, 2001) signaling pathways. Similar to the muscarinic agonists, nicotinic agonists such as nicotine or DMXB (selective α7 nAChR agonist) have memory-enhancing activity and are involved in neuronal survival (Kihara *et al.*, 2001).

These data suggest that ERKs may serve as a convergence point for different signaling pathways in mature neurons to produce plasticity. Activation of neurotrophin receptors (Lu and Figurov, 1997), AChR receptors (Auerbach and Segal, 1996), and NMDA receptors (Collingridge *et al.*, 1983) can all induce LTP and ERK1,2 activation. Thus, fast (e.g. glutamaergic) and modulatory (e.g. cholinergic) neurotransmission, both necessary for normal learning and memory, may converge on ERKs in a given neuron, and therefore expanding the repertoire of responses available to a cell exposed to a given extracellular signal.

2.   AP-1 transcription factor complexes in the CNS

AP-1 (activator protein 1) is a collection of dimers composed of jun and fos families of bZIP (basic region-leucine zipper) DNA binding proteins. These dimers bind to a common cis acting element known as the TRE (TPA response element) or the AP-1 site (general sequence TGACTCA) in gene promoters. While many of the AP-1 factors are transcriptional activators, certain AP-1 complexes can function as transcriptional repressors (Krzanowski, 1995).

The c-fos and fos-related antigens (fra) family includes fra-1, fra-2 and FosB, whereas c-Jun, JunB and JunD comprise the Jun-related factors. All are in general highly inducible, with the exception of JunD which is constitutively expressed in some tissues and is induced a few fold (Schlingenspien *et al.*, 1994; Minden and Karin, 1997).

A functional DNA binding complex is formed when fos (or fra) dimerizes with a Jun-related factor; however, the Jun proteins can heterodimerize with other factors, such as CREB or ATF, or homodimerize with themselves. The composition of these complexes determines the specificity of cellular response. There is also evidence of the existence of proteins that can bind c-Jun and JunD, but not JunB, and thus stabilize specific Jun complexes (Claret *et al.*, 1996). Moreover, whereas c-Jun stimulates the transcription of c-Jun gene, JunB inhibits it (Tong *et al.*, 1998). c-Jun transcription is rapidly stimulated, in the absence of *de novo* protein synthesis, following exposure to a variety of extracellular stimuli, including growth factors, proinflammatory cytokines and UV irradiation (Minden and Karin, 1997). Phosphorylation of Ser 63 and Ser73 residues of c-Jun by JNK appears to be critical for stimulating c-Jun transcriptional activity. Of the different Jun proteins, only c-Jun is an efficient JNK substrate whereas JunB and JunD phosphorylation is much less efficient (Minden and Karin, 1997). In addition to c-Jun, the JNKs also phosphorylate and activate other transcription factors such as ATF-2 (also known as CRE-BP1) on Thr 69 and 71 residues (Gupta *et al.*, 1995).ATF-2 can be phosphorylated/activated also by p38 MAPK (Angel *et al.*, 1991). Interestingly, c-Jun and ATF-2 dimerize and cooperatively stimulate the expression of c-Jun gene by binding to the non-consensus TRE in the c-Jun promoter (van Dam *et al.*, 1993) (Fig. 2). JNK pathway can also lead to increased expression of another component of AP-1, c-fos.

c-fos gene induction occurs in a rapid and transient manner in response to a variety of extracellular stimuli in various cell types. Growth factor stimulation of the c-fos gene was shown to be mediated at least in part by phosphorylation of Elk-1 by ERKs (Marais *et al.*, 1993). However, c-fos transcription is also enhanced by other stimuli (IL-1, UV irradiation, TNFα) or enzymes (MEKK) which are poor activators of ERKs. It has been shown that JNK and p38 MAPK phosphorylate Elk-1 on the same major sites as do the ERKs, stimulating its transcriptional activity (Minden and Karin, 1997). Thus, while induction of c-fos gene by growth factors is primarily mediated by ERKs, the SAPKs (JNK and p38 MAPK) are most likely responsible for c-fos induction in response to cellular stresses and cytokines. Interestingly, the newly synthesized c-fos protein plays a role in repression of its own promoter (Wilson and Treisman, 1988; Schonthal *et al.*, 1989) and the ERK/RSK signal transduction pathway participates in both the transcriptional activation and repression of c-fos gene expression (Chen *et al.*, 1993).

The family of CREB transcription factors recognizes the CRE (general sequence TGAGCTCA) in gene promoters and links cAMP signal transduction to the nucleus (Brindle and Montminy, 1992). It includes CREB, ATF 1 to 5, CRE modulators (CREM) and CREB-327. Except for CREM, these factors are constitutively expressed but activate transcription only after being phosphorylated on specific amino acid residues. Certain genes may have both AP-1 and CRE sites in their promoters, thereby providing a complex regulation. Although CREB was first identified as a mediator of gene expression that occurs in response to increased concentrations of cAMP, CREB also regulates the cellular response to growth factor stimulation (Ginty *et al.*, 1994). Phosphorylation of CREB at Ser133 by RSK, Akt or MAPKAP2 [and its subsequent association with coactivator protein CBP (CREB-binding protein) or p300], triggers the activation of CREB transcription activity and, among others, is critical for growth factor induction of c-fos transcription (Ginty *et al.*, 1994).

AP-1 DNA binding activity was increased in every region of the rat brain during the first postnatal week and returns to low basal levels by the third week (Pennypacker *et al.*, 1995). The AP-1 complex (seen between postnatal weeks 1-3) is composed of c-Jun-CREB dimer and regulates developmentally important genes such as glial fibrillary acidic protein (GFAP), which is involved in astroglial differentiation and growth associate protein 43 (GAP43), which is associated with formation of neuronal circuitry. In the adult rat, the AP-1 activity remains relatively elevated in the olfactory bulb and cerebellum, but the composition of the complexes differ for the two brain regions: fra antigens are predominant in the olfactory bulb (particularly in the granule cells) while c-Jun-CREB dimers are predominant in the cerebellum (Pennypacker *et al.*, 1995). This difference may be due to cellular activities that are region-specific, such as synaptogenesis and/or regeneration in the olfactory bulb and differentiation/proliferation in the cerebellum. There is good evidence that c-Jun is required for developmental programmed cell death (Ham *et al.*, 1995). Expression of c-Jun is increased in NGF-deprived rat sympathetic neurons undergoing apoptosis. Microinjection of antibodies specific for c-Jun or c-Fos protect these NGF-deprived neurons from death whereas anti-JunB and anti-JunD do not (Estus *et al.*, 1994). In the rodent brain, JunD is fairly ubiquitous whereas c-fos, JunB, c-Jun and Fos B are found in a more restrictive an differentiated fashion, especially in the hippocampus, where c-Jun is predominantly localized to the dentate gyrus and JunB to the CA-1 area of the hippocampus (Herdegen *et al.,* 1995). Significantly, AD pathological manifestations are associated with activation of JNK and its substrates c-Jun and ATF-2 (Yamada *et al.*, 1997; Morishima *et al.*, 2001; Shoji *et al.*, 2001). In the CNS, c-fos was first reported to be induced in the hippocampus due to seizure activity (Sonnenberg *et al.*, 1989) and its expression increases after traumatic brain injury (Yang *et al.*, 1994) and in neurodegenerative diseases (Smeyne *et al.*, 1993; Pennypacker *et al.*, 1994). Sustained and/or elevated c-fos expression is associated with neuronal apoptosis and developmental failures both *in vitro* and *in vivo*

(Ruther et al., 1987; Smeyne et al., 1993). Both c-Jun and JunB dramatically increase after seizure activity while JunD is only modestly increased (Krzanowski, 1995). Ischemic brain injury is also associated with increases in c-fos, Jun B, c-Jun and ATF-2 (Tong et al., 1998; Walton et al., 1998), suggesting that these proteins can be used as markers of neuronal damage. The link between transcriptional regulation of AP-1 and DNA repair (which is impaired in apoptosis) is Ref-1, a bifunctional protein. Loss of Ref-1 protein expression was shown to precede DNA fragmentation in apoptotic neurons (Walton et al., 1997). Significant to the role that AP-1 transcription factors may play in the CNS disease, a strong correlation has been established between amyloid beta (Aβ)-induced neuronal death (occurs in AD) and altered gene expression involving c-Jun-Fos heterodimers (Estus et al., 1997). Aβ-treatment of cortical neurons induced expression of c-Jun, c-fos, fos B and JunB and the AP-1 complex (constituted by c-Jun-c-fos) was responsible for induction of target genes such as transin (which encodes an extracellular matrix protease (Estus et al., 1997). Moreover, c-Jun and c-Fos expression is also induced in vivo, in tangle-bearing neurons in AD-affected brains (Anderson et al., 1994).

By contrast, CREB phosphorylation and activation occur as a neuronal response to neurotrophins and are important for the survival of neurons in a number of different conditions (Finkbeiner et al., 1997). Moreover, CREB activation couples alterations in gene expression not only with survival but also with neuronal activity, since it has been implicated in synaptic plasticity and cognitive functions (Bailey et al., 1996; Silva et al., 1998). ERKs play a critical role in CREB phosphorylation/activation in the hippocampus by either triggering it (via the intervening kinase RSK) or acting as an obligatory intermediate or conduit in PKA and PKC activation of CREB (Roberson *et al.* 1999). Moreover, the CaMK pathway will most likely recruit the ERK pathway in response to strong Ca $^{2+}$-mobilizing stimuli thus leading to a more prolonged increase in CREB activation (Wu *et al.*, 2001). Also, a variety of neurotransmitter receptors (metabotropic glutamate receptors, mAChR, nAChR, dopamine receptors, and β-adrenergic receptors) converge on ERKs via the PKA/ PKC cascades (Roberson *et al.*, 1999) or calcium-dependent NO production (Sasaki *et al.*, 2000), as illustrated in Fig. 3, providing a complex interplay in regulating CREB-mediated gene transcription and activity-dependent functions.

In conclusion, AP-1 and CREB transcription factors are involved in a variety of neuronal activities, including proliferation/differentiation, survival or activity-dependent processes. The composition of AP-1 determines the specific cellular response to extracellular stimuli.

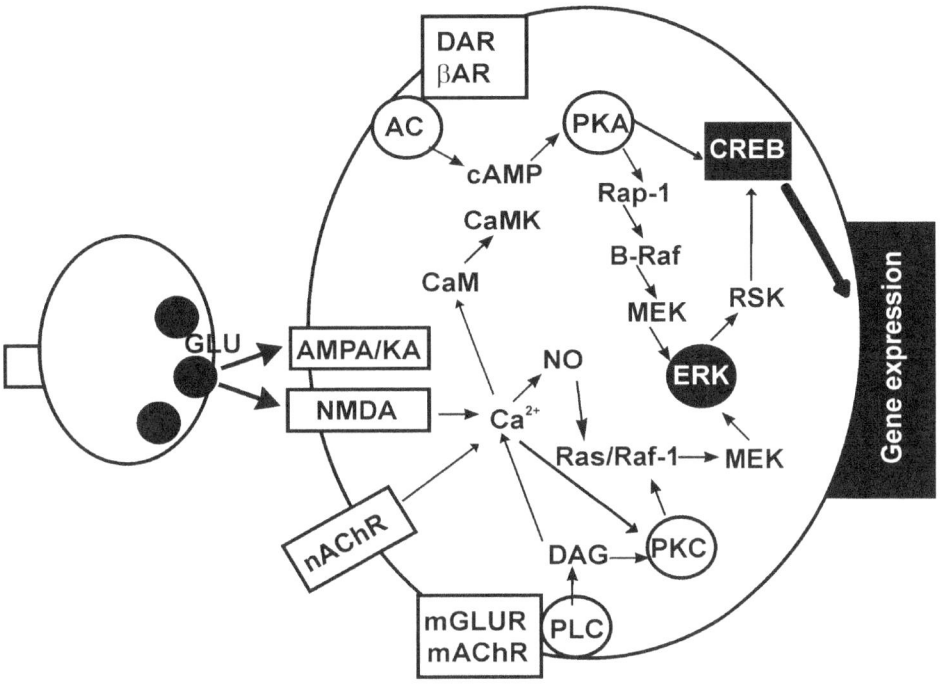

Fig. 3.   CREB as a convergence point for signaling pathways involved in

hippocampal synaptic plasticity. The MEK/ERK pathway serves as a conduit for

signaling from glutamate (AMPA/KA, NMDA, mGLU) or neurotransmitter receptors

(DAR, βAR, mAChR, nAChR) leading to CREB activation. The convergence on final

effectors of distinct pathways may provide a fail-safe mechanism (the failure of one may

be compensated by another) or/and signal amplification by temporally spaced stimuli.

GLU, glutamate; *NO*, nitric oxide; *AMPA/KA*, α-amino-3-hydroxy-5-methyl-4-isoxazole

propionic acid/kainic acid receptor; *NMDA*, N-methyl-D-aspartate receptor; *mGLU*,

metabotropic glutamate receptor; *mAChR*, muscarinic acetylcholine receptor; *nAChR*,

nicotinic acetylcholine receptor; *DAG*, diacylglycerol; *PKC*, protein kinase C; *CaM*,

calmodulin; *CaMK*, calmodulin kinase II or IV; *PLC,* phospholipase C; *AC*, adenylate

cyclase; *DAR*, dopamine receptor D1/D5; *βAR*, beta-adrenergic receptor; *PKA*, protein

kinase A; *RSK*, ribosomal S6 kinase, also known as CREB kinase.

B. **Apoptosis in development and disease**

Cell death can be simplistically categorized as either necrotic (i.e. pathological and accidental) or apoptotic (i.e. physiological and planned). Apoptosis or programmed cell death is an important cellular response that serves to control the cell number by removing damaged, unwanted or virus-infected cells. It describes the controlled destruction of single cells after activation of the death program. All mammalian cells constitutively express the proteins required to execute this death program (Jacobson *et al.*, 1994). There are 3 distinct phases along the apoptotic path : i) commitment phase, in which the cell becomes committed to die, and depends on the nature and duration of the stimulus; ii) execution phase, in which the major morphological and structural changes associated with apoptosis occur, and iii) clearance phase when apoptotic bodies are removed by professional phagocytes or neighboring cells (Johnson-Webb *et al.*, 1997). Mitochondria play a key role on the path to irreversible commitment to apoptosis, as illustrated in Fig. 4.

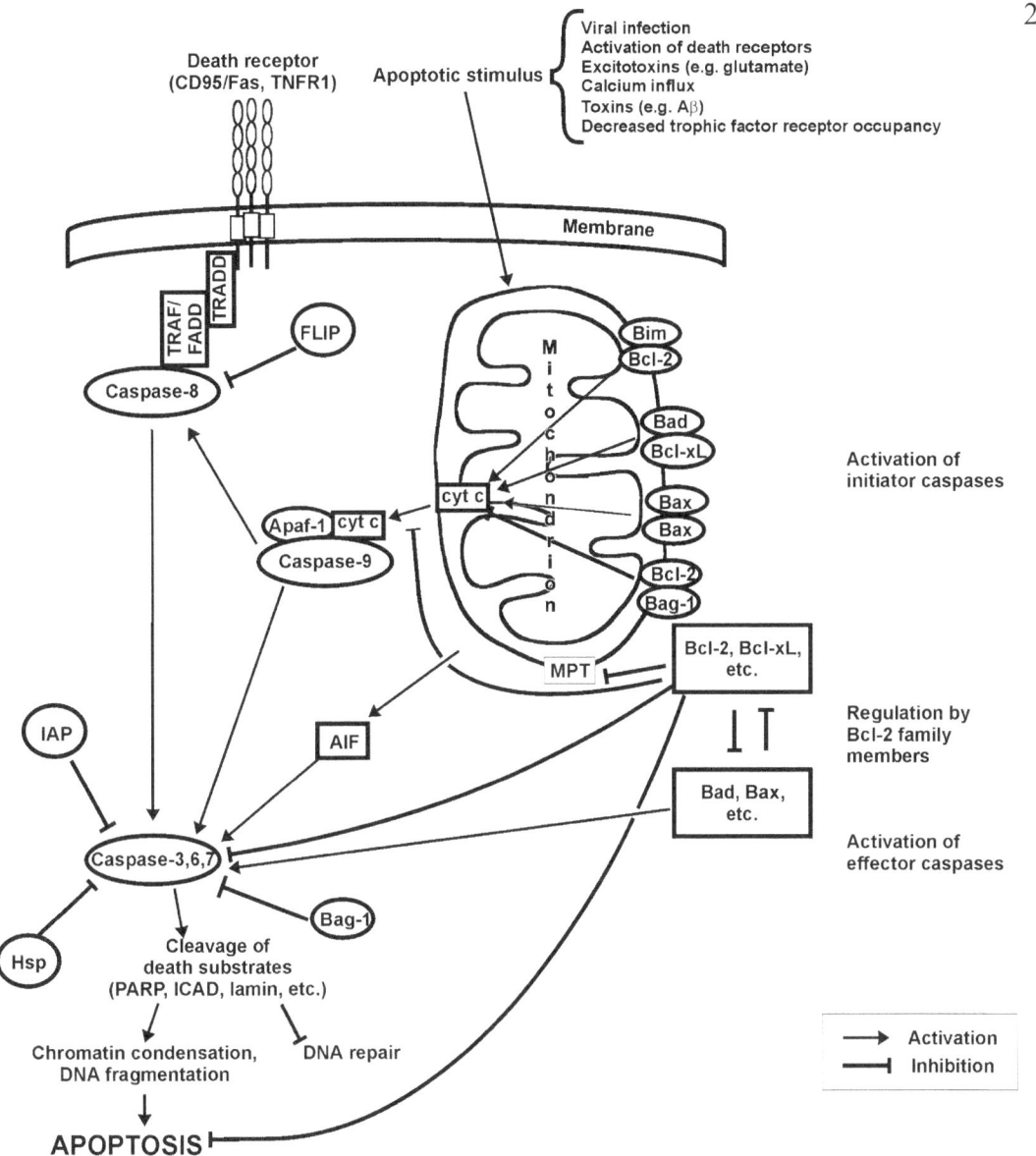

Fig. 4. **Apoptotic pathways.** Apoptotic stimuli may directly act on mitochondria

inducing apoptosis-specific mitochondrial changes such as release of cytochrome c (cyt c)

and apoptosis inducing factor (AIF), mitochondrial permeability transition (MPT) or

changes in the dimerization status or expression levels of Bcl-2 family of proteins (such as Bcl-2, Bcl-xL, Bad, Bax, Bim, etc.). Released cyt c forms a complex with Apaf-1 and caspase-9 leading to activation of proteolytic caspase cascade and apoptosis. Activation of "death receptors" (CD95/Fas, TNFR1) by ligand binding (Fas, TNF) leads to the recruitment of adaptor proteins (such as TRAF, FADD), activation of caspase-8 and subsequent activation of proteolytic caspase cascade leading to apoptosis.

Apoptotic stimuli may directly influence the mitochondrial functions by inducing mitochondrial alterations such as: i) mitochondrial permeability transition (MPT), which is characterized by the opening of pores in the inner mitochondrial membrane leading to a fall of the membrane potential and the arrest of ATP synthesis, and ii) release of cytochrome c (cyt c) or apoptosis-inducing factor (AIF). Cyt c and AIF are generally required for the activation of the cell death executioners, caspases (Mignotte and Vayssiere, 1998). Caspases belong to a 12 member family and are cysteine proteases with aspartate specificity that are activated by proteolytic cleavage of zymogens (pro-caspases, composed of three domains: an N-terminal pro-domain, and the p20 and p10 domains). Most caspases are activated by proteolytic cleavage of the zymogen between the p20 and p10 domains, and usually also between the prodomain and the p20 domain (Hengartner, 2000). AIF, which appears to be a protease, can directly activate the caspase cascade. On the other hand, cyt c released from mitochondria forms a complex with pro-caspase-9 and Apaf-1 - which in this context acts as a potential regulatory subunit of the holoenzyme (caspase-9). This complex is often referred to as the "apoptosome" and its formation leads to activation of caspase-9 ("initiator caspase") by means of conformational change (Hengartner, 2000). Caspase-9 triggers the activation of a caspase cascade involving caspases-3, 6 and 7 ("effector" caspases) through proteolytic cleavage of the respective zymogens. Caspase-3 cleaves proteins involved in DNA repair and replication, such as poly (ADP-ribose) polymerase (PARP) (Lazebnik *et al.*, 1994) and inhibitor of caspase activated DN-ase (ICAD) as well as structural proteins, such as lamin and fodrin (Janicke *et al.*, 1998). The effector caspases are ultimately responsible for the morphological and biochemical changes associated with apoptosis which include chromatin condensation and DNA fragmentation as well as inhibition of DNA repair. Another apoptotic pathway (Fig. 4) is initiated by binding of Fas or TNF to their respective receptors (Fas/CD95 and TNFR1) called "death receptors". They contain cytoplasmic domains called "death domains", which serve to anchor an intracellular protein called TRADD that serves as an adaptor for binding TRAF (TNF receptor-associated factor) or FADD (Fas-associated death domain) proteins (for TNFR1 and Fas/CD95 receptor respectively). FADD also carries a so-called death effector domain (DED) which is involved in recruitment and activation of DED-containing procaspase-8 (also known as FLICE). Another member of the family, caspase-10 may also participate in this pathway as an "initiator" caspase (Hengartner, 2000) . The subsequent step is similar to the pathway initiated by caspase-9 and involves the activation of "effector" caspases.

Apoptotic pathways are regulated by a family of related proteins with Bcl-2 as a prototypic member. Studies with the Bcl-2 oncogene first indicated that apoptosis may play an important role in oncogenesis. Thus, Bcl-2 transforming activity rests on its ability to inhibit apoptosis rather than cause cell proliferation (Korsmeyer, 1992). The mammalian Bcl-2 related family of proteins (reviewed by Strasser *et al.*, 1997) (Table 1) are characterized by the presence of Bcl-2 homology domains and the capacity to modulate apoptosis.

Table 1

Mammalian Bcl-2 family members		
Anti-apoptotic (1st class)	**Pro-apoptotic (2nd class)**	
Bcl-2	Bcl-xS	Hrk
Bcl-xL	Bax	Bin
Bcl-w	Bak	Bad
Mcl-1	Bok	Noxa
A1	Bik	Diva
Nr-13	Bid	

They can be functionally divided into two classes. The first class includes the prototypic family member, Bcl-2 and related anti-apoptotic proteins (such as Bcl-xL, A1, etc.). The second class includes the pro-apoptotic members of the family such as Bad and Bax. As illustrated in Fig. 4, pro- and anti-apoptotic family members heterodimerize (e.g. Bcl-

2/Bax, Bcl-xL/Bad, Bcl-2/Bad, A1/Bax) and the balance between the two classes determines, at least in part, the susceptibility to apoptosis (Tsujimoto, 1998). The mechanism of action of the Bcl-2 family of proteins is unclear. Since Bcl-2 and Bcl-xL are mainly localized at the mitochondrial membrane, it was originally postulated that their main activity involves inhibition of mitochondrial dysfunction by interference with MPT and cyt c release from mitochondria (Chittenden, 1998) possibly through their pore-forming activity. Pro-apoptotic Bcl-2-like proteins are also located at the mitochondrial membrane and they stimulate cyt c release either directly (Bax-Bax homodimers) or indirectly, through heterodimerization with anti-apoptotic members, by relieving the blockade imposed by Bcl-2 or Bcl-xL (e.g. Bim-Bcl-2 and Bad-Bcl-xL heterodimers). However, apart from their mitochondrial-related functions, the anti-apoptotic Bcl-2 proteins are also able to interact with pro-caspases [e.g. pro-caspases-8 and pro-caspase-9 (via Apaf-1)] blocking their activation (Hu *et al.*, 1998; Tsujimoto, 1998). Moreover, caspase-8 was shown to cleave Bid [a pro-apoptotic family member (Wang *et al.*, 1996)]. thereby activating its pro-apoptotic function which is associated with mitochondrial dysfunction (Li *et al.*, 1998). Cleavage of Bcl-2-related proteins by caspases is not limited to the pro-apoptotic family members. During infection with Sindbis virus, Bcl-xL is also cleaved by caspases, a process associated with apoptosis (Clem *et al.*, 1998). In the same fashion, Bcl-2 was shown to be cleaved by caspases releasing a pro-apoptotic cleavage product (Cheng *et al.*, 1997).

Considering that Bcl-2 prevents many, but not all, forms of apoptotic cell death (Tsujimoto, 1998) it is possible to suggest that there are multiple independent intracellular mechanisms of apoptosis involving proteins that differentially regulate Bcl-2 function (Boise *et al.*, 1993). Such a protein is the Bcl-2 associated athanogene (Bag-1). Bag-1 (originally named RAP46) is expressed as different isoforms generated by alternative translation initiation (Packham *et al.*, 1997). The small isoforms (32-36 kDa, Bag-1S) are cytoplasmic, whereas the larger isoforms (46-50 kDa, Bag-1M, Bag-1L) have a nuclear localization (Packham *et al.*, 1997; Takayama *et al.*, 1998). Bag-1 interacts physically and functionally with Bcl-2 to cooperatively interfere with the apoptotic cascade at the level of caspase activation (Takayama *et al.*, 1995; Bardelli *et al.*, 1996; Schulz *et al.*, 1997). Not only does it enhance the anti-apoptotic activity of Bcl-2, but it also inhibits apoptosis by itself (Takayama *et al.*, 1995). Bag-1 collaborates with Bcl-2 in suppressing apoptosis induced by staurosporine, growth factor withdrawal, chemotherapeutic drugs or CD95/Fas receptor activation (Takayama *et al.*, 1995; Clevenger *et al.*, 1997; Schulz *et al.*, 1997; Terada *et al.*, 1997). This effect may be due to a stabilization of the Bcl-2 protein in the Bag-1-Bcl-2 complex, thereby leading to a change in the balance of pro- to anti-apoptotic protein family toward the latter. Moreover, due to its presence in the mitochondria (Takayama *et al.*, 1998), Bag-1 may directly inhibit MPT and cyt c release. Recent reports described the interaction of Bag-1 with several signaling proteins (such as Raf-1, HGF and PDGF), hormone receptors or heat shock protein 70, suggesting that it has a general facilitatory role in signal transduction pathways involved in regulation of cell survival (Takayama *et al.*, 1997). Bag-1 was also shown to bind SIAH-1 (which can induce apoptosis by promoting upregulation of p53), and inhibit p53-mediated apoptosis (Matsuzawa *et al.*, 1998). Bag-1 has been shown to, by itself, inhibit or delay apoptosis caused by growth factor withdrawal, heat shock, and p53 (Takayama *et al.*, 1995; Clevenger *et al.*, 1997; Durten-van Oorschot *et al.*, 1997; Takayama *et al.*, 1997). Bag-1 may also play a role in normal CNS function, as evidence by its presence in the normal but not in the ischemic-injured brain (Hayashi *et al.*, 2000). The DNA damage was preceded by loss of Bag-1 expression, which most likely disturbed the balance between pro- and anti-apoptotic protein expression, caused disturbed molecular chaperone activity or neurotrophin induced survival signal and thus resulted in neuronal cell death (Hayashi *et al.*, 2000).

Other families of apoptosis modulators are: FLIPs (FLICE-inhibitory proteins), IAPs (inhibitors of apoptosis) and heat shock proteins (Hsp).

FLIP contains two DEDs and binds to the CD95-FADD complex leading to inhibition of caspase-8, formerly known as FLICE (Krammer, 2000). It was suggested that cellular FLIP (c-FLIP), also known as Casper, Cash, I-FLICE or MRIT (Dorstyn *et al.*, 1998) can be either pro- or anti-apoptotic depending on the cell type (Goltsev *et al.*, 1997). However, recent data with cells from c-FLIP-deficient mice support the role of c-FLIP as an anti-apoptotic molecule (Yeh *et al.*, 2000).

The IAP family has five homologous members (NAIP, cIAP1/HIAP2, cIAP2/HIAP1, hILP/XIAP, and survivin that act by inhibiting activated caspases (Jaattela, 1999). hIAP1 and hIAP2 associate with TRAFs, thereby interfering with signal transduction through the death receptors (Hardwick, 1997). hILP/XIAP fails to bind TRAFs but reportedly binds and inhibits caspase-3 and 7 (Deveraux *et al.*, 1997). Mutations in NAIP were found associated with spinal muscular atrophy (SMA) and ALS

(Levine *et al.*, 1993), which are motor neuron diseases that result from the loss of motor neurons due to excessive apoptosis. It has been proposed that NAIP is a tissue (motor neuron)-specific anti-apoptotic IAP gene whose inactivation could result in neurological deficits.

Heat shock proteins (Hsps) are proteins synthesized in cells in response to an elevation in temperature. Several Hsps have been shown to have anti-apoptotic activity, believed to be due to their chaperone activity, in a variety of paradigms. For example, Hsp70 can mediate cell protection from heat stress by inhibiting the effector caspases (Mosser *et al.*, 1997). However, it can also mediate protection even in the presence of activated effector caspases (Jaattela *et al.*, 1998), suggesting a more complex mechanism of action. The putative mechanisms of this protective activity of Hsp70 are believed to be due to its interaction/inhibition of pro-apoptotic proteins such as p53 or c-myc (Sturzbecher *et al.*, 1988; Koskinen *et al.*, 1991). Moreover, Hsp70 binds to the anti-apoptotic protein Bag-1, found also in association with Bcl-2 (Takayama *et al.*, 1998). Bag-1 binds to the ATP-ase domain of Hsp70 and inhibits its chaperone activity by blocking the binding of a cochaperone, Hip. The relevance of this binding is, however, unclear. Theoretically, Bag-1 could increase the binding of some pro-apoptotic proteins to Hsp70 by inhibiting their ATP-dependent release from Hsp70 (Jaattela, 1999). Hsps have been shown to have anti-apoptotic activity in neurons. Hsps 27, 70 or 90 can protect dorsal root ganglion neurons from thermal or ischemic stress by a mechanism not yet identified (Uney *et al.*, 1993; Amin *et al.*, 1996). It is noteworthy that the interpretation of their anti-apoptotic activity as therapeutically beneficial is complicated by the finding that these proteins are overexpressed in cerebral ischemia *in vivo* (Wagstaff *et al.*, 1996), which complicates

The third phase of the apoptotic process is the phagocytosis and clearance of apoptotic cells. An apoptotic cell displays a number of markers, such as the exposure of phosphatidylserine residues (normally restricted to the inner leaflet of the dying cell), or changes in surface sugars (detected by phagocyte lectins) that allow phagocytes to detect and mediate the engulfment of dying cells (Savill and Fadok, 2000). Clearance is achieved mostly by macrophages, although the neighboring cells may also participate. Macrophages are essential for both the clearance of apoptotic cells generated in an injured tissue and for host defense against infection with microorganisms. The uptake of apoptotic cells actively suppresses the secretion from activated macrophages of pro-inflammatory cytokines, such as TNFα (which normally occurs during clearance of microorganisms), thereby uncoupling the apoptotic process from inflammatory responses and preventing secondary necrosis of apoptotic cells (Voll *et al.*, 1997).

Necrotic cell death appears traumatically and may be described as an unregulated process that occurs when the cell is exposed to extremely unphysiological stimuli. Necrosis starts with the loss of control over the cellular homeostasis, leading to swelling of organelles and the whole cell and terminating in cell lysis. Because of the ultimate destruction of the cell membrane, the cytoplasmic content, including lysosomal enzymes, is released into the extracellular space. Therefore, necrotic cell death *in vivo* is often accompanied by extensive tissue damage, resulting in an intense inflammatory response. Apoptosis and necrosis often coexist in many diseases of the CNS and also in neurons stimulated *in vitro* by excitotoxins (Ankarcrona *et al.*, 1995; Choi, 1995). For example, they occur simultaneously after induction of hypoxia in PC12 cells or upon potassium withdrawal from cerebellar granule cell cultures (Ishitani *et al.*, 1997; Villalba *et al.*, 1997). Alternatively, they may occur in a temporal sequence. For example, a short exposure of neurons to glutamate kills many of them rapidly by necrosis, but the surviving cells eventually undergo delayed apoptosis (Ankarcrona *et al.*, 1995). On the other hand, necrosis sometimes follows apoptosis, especially in experimental systems that lack the phagocytic means to clear rapidly all apoptotic cell (Leist *et al.*, 1995; Portera-Cailliau *et al.*, 1995). Apoptosis and necrosis may also follow a certain spatial distribution within a tissue. For example, in cerebral ischemia, necrosis is prevalent in the core region whereas apoptosis is more prevalent towards the border regions, where energy depletion and excitotoxic stimulation are less severe and prolonged (Arenas and Persson, 1994; Charriaut-Marlangue *et al.*, 1996). There seem to be a common view that apoptosis and necrosis are the extremes of a continuum of possible types of cell death, and that, by increasing the intensity (exposure time or concentration) of the insult, cell death can change its shape from apoptotic to necrotic (Leist and Nicotera, 1998; Nicotera and Leist, 1997).

It is now clear that apoptosis or programmed cell death is an essential component of animal development, important for establishment and, in vertebrates at least, maintenance of tissue architecture. Apoptosis also acts as part of a quality-control and repair mechanism that contributes to the high level of plasticity during development by compensating for many genetic or stochastic developmental errors. For example, apoptosis is one of the major processes that act to shape and define the final form of the functional mature nervous system, by removing neurons produced in excess or inappropriate connections (Oppenheim, 1991). Apoptotic death occurring in development is therefore a normal, physiological process. However, since neurons become postmitotic and, therefore, irreplaceable, the occurrence of apoptosis outside the circumscribed developmental period, is considered a pathological process. The decision of whether a neuron dies or lives during development is determined primarily by the availability of neurotrophic factors (such as NGF), even though other influences (e.g. afferent input) may also be important (Johnson *et al.*, 1993). NGF can prevent neuronal death occurring during development and after target removal or axotomy. Similarly, administration of NGF-neutralizing antibodies to immature animals results in massive death of specific neuronal populations, suggesting that NGF is required for survival (Levi-Montalcini and Booker, 1960).

In mature organisms, disruption of the cell death regulation, resulting in too little or too much apoptosis leads to pathological manifestations such as cancer and neurological disorders.

Defects in apoptosis regulation are common in cancer cells. Such defects (leading to resistance to apoptosis) may play a role in tumor initiation (since apoptosis normally eliminates cells with damaged DNA that have an increased malignant potential) and metastasis (by allowing the cancerous cells to survive the transit in the bloodstream and to grow in ectopic tissue sites lacking the otherwise required survival factors). Among others, Bcl-2, Hsp27, Hsp70 and survivin are cancer-associated anti-apoptotic proteins that are currently targets in cancer therapy. The reduction in the respective protein levels by an antisense approach, renders tumor cells more sensitive to cancer drugs (Jaattela, 1999). Moreover, the major mode of action of chemotherapeutic drugs and radiotherapy is to induce apoptosis in cancer cells, most frequently by DNA-damage-induced activation of p53 (Watson and Lowenstein, 1998). This explains the clinical observation that tumors with mutant p53 are resistant to chemo- or radiotherapy, since p53-dependent apoptosis cannot be activated by DNA damage in these tumors (Lens *et al.*, 1997; Berns *et al.*, 1998).

By contrast to the situation in cancer, defects in regulation of apoptosis leading to excessive cell death are associated with pathological manifestations of neurological disorders. Among the common causes leading to excessive apoptosis are: excitotoxicity (high glutamate concentrations or deregulated generation of NO, leading to production of reactive oxygen species/oxidative stress) and loss of target-derived neurotrophins. Apoptosis is involved in the etiology of neurodegenerative disorders [such as AD, PD and ALS] and acute brain injury (such as ischemia/hypoxia and trauma). The incidence of AD increases exponentially with age and is especially seen in the population over 65 years of age with prevalence ranging from 28 to 47 % (Martin, 1996). Hallmarks of AD include the presence of senile plaques consisting of extracellular deposits of amyloid beta ($A\beta$) protein and selective loss of neurons due to apoptosis, especially in regions of the hippocampus that are associated with memory and learning impairing cognitive and memory processes (Honig and Rosenberg, 2000). Moreover, cholinergic neurons of the basal forebrain are affected early in the disease due to the loss of target-derived neurotrophic factors. The mechanisms of Aβ neurotoxicity include induction of oxidative stress, elevation of intracellular Ca^{2+} concentration, activation of caspases and DNA fragmentation (Honig and Rosenberg, 2000; Junying and Yankner, 2000). Down syndrome (DS) is the most common clinical syndrome associated with mental handicap. It occurs in about 1 of 1000 live births and accounts for about 15 % of the total mentally handicapped population. DS is associated with a full trisomy of chromosome 21, due to chromosomal non-disjunction during meiosis. There are common pathological mechanisms between AD and DS. Similar to AD patients, elderly persons with DS have a lower number of pyramidal and non-pyramidal nerve cells in the temporal cortex, hippocampus and entorhinal cortex, than similarly aged normal people in the general population, suggesting that the genetic defect in DS confers increased vulnerability to neurodegeneration (Sawa, 1999). Notwithstanding DS patients develop AD by their fifth decade of life. Cultured neurons from both patients and animal models (such as the trisomy 16 mouse) are more vulnerable to apoptosis, thus providing evidence for this type of neuronal cell death in DS (Richards *et al.*, 1991; Sawa, 1999). Increased expression of several apoptosis-related genes (p53, Fas ligand, ratio of Bax to Bcl-2) has been reported in cultured DS neurons or DS brains. Possible mechanisms include p53 activation, overproduction of reactive oxygen species and toxicity of Aβ (also detected in DS brains, similar to AD) (Sawa, 1999). PD is a degenerative disease characterized by selective neuronal loss in the substantia nigra, locus coeruleus and dorsal motor nucleus of the vagus. Its prevalence is 1.4 % in those over 55 years ad 4.3 % in 65 and older (de Rijk *et al.*, 1995). In PD, abnormal deposits of the protein α-synuclein, oxidative stress and loss of target (striatum)-derived growth factors such as glial-derived neurotrophic factor, are associated with neuronal cell apoptosis (Honig and Rosenberg, 2000). ALS is a progressive motor disease characterized by the degeneration of motor neurons in the spinal cord and brain, leading to paralysis. Familial ALS is associated with mutations in the gene encoding superoxide dismutase 1 (SOD-1) (Rosen *et al.*, 1993). Mutant SOD-1 can form intranuclear aggregates and induce oxidative stress (reminiscent of the pathological mechanism in AD) (Cleveland, 1999). Moreover, activated caspase-1 and caspase-3 were detected in spinal cords of ALS patients and mutant SOD-1 transgenic mice (Li *et al.*, 2000). All these neurodegenerative disorders display their pathological manifestations (including neuronal loss due to apoptosis) over a period of many years, in contrast to the fast apoptosis that occurs in acute brain injury (stroke, trauma and hypoxia-ischemia). In these disorders, the initial insult to the brain causes necrosis but also a delayed "secondary" injury in a "penumbra zone", initially spared and surrounding an area of most severe damage. The mechanism of cell death in these penumbra areas is apoptotic in nature and is amenable to anti-apoptotic therapeutic strategies that will minimize the brain damage (Honig and Rosenberg, 2000).

A summary of the putative apoptotic pathways in neurons is presented in Fig. 5.

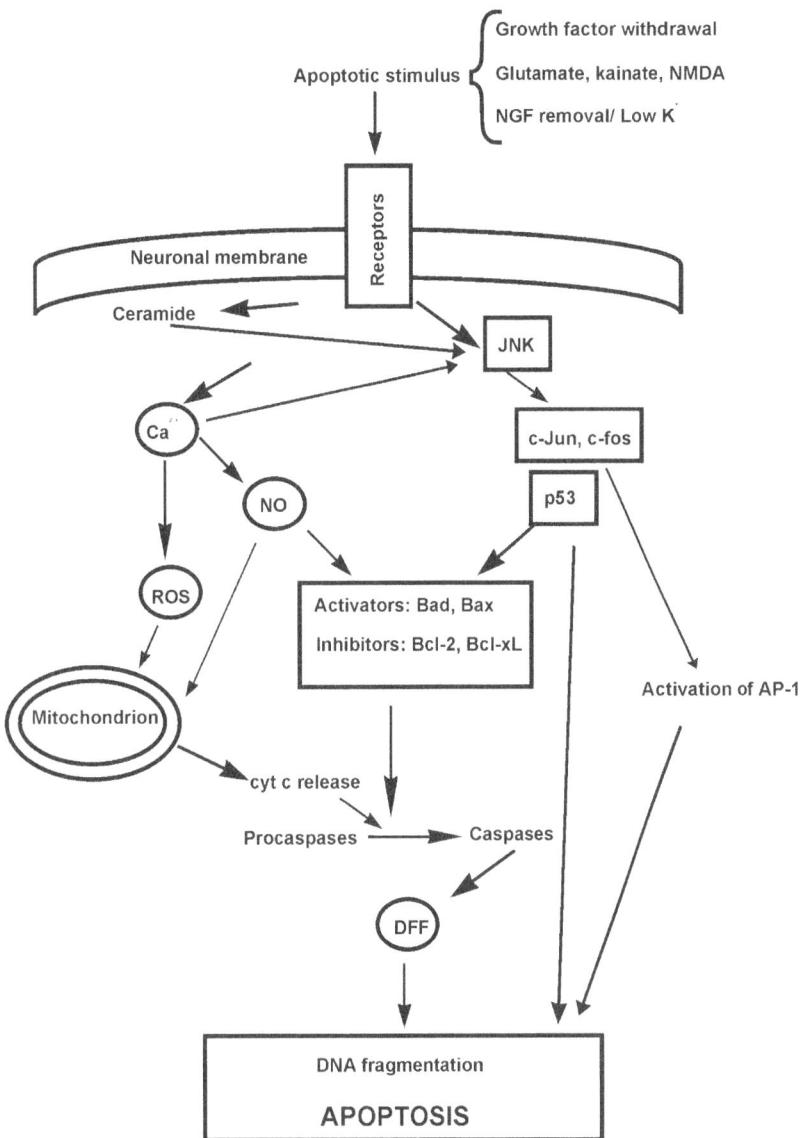

Fig. 5. **Apoptotic pathways in the nervous system.** Activation of p75NTR,

glutamate, etc. receptors leads to activation of pro-apoptotic JNK pathway and activation

of AP-1 complex, mitochondrial dysfunctions due to ROS and increased expression of

pro-apoptotic members of Bcl-2 family. Subsequent caspase activation triggers DNA

fragmentation and apoptosis. *NMDA*, –methyl-D-aspartate; *JNK*, c-Jun-N-terminal

kinase; *NO*, nitric oxide; *ROS*, reactive oxygen species; *cyt c*, cytochrome c; *DFF*, DNA

fragmentation factor.

1. Apoptosis modulators and signaling proteins

Signal transduction pathways lie upstream of the cell apoptotic machinery and modulate the apoptotic response. In this respect, Bcl-2 protein is able to interact with signaling proteins. It associates with Raf-1 targeting this kinase to the mitochondrial membrane where it phosphorylates the pro-apoptotic Bcl-2-like protein Bad suppressing its death-inducing function (Wang *et al.*, 1996) (Fig. 6). The phosphorylated Bad is unable to neutralize the anti-apoptotic functions of Bcl-xL or Bcl-2 by heterodimerization (Fig. 4 and Fig. 6), instead it is sequestered in the cytoplasm bound by a phosphoserine-binding protein, 14-3-3 (Zha *et al.*, 1996) which was previously shown to interact with Raf-1 enhancing its kinase activity (Freed *et al.*, 1994). Bad may also be phosphorylated by Akt (Datta *et al.*, 1997) or RSK 1 (Shimamura *et al.*, 2000) with the same end-result, survival and inhibition of apoptosis, while its dephosphorylation by calcineurin or phosphatase 1α (PP1α) is associated with apoptosis (Wang *et al.*, 1999).

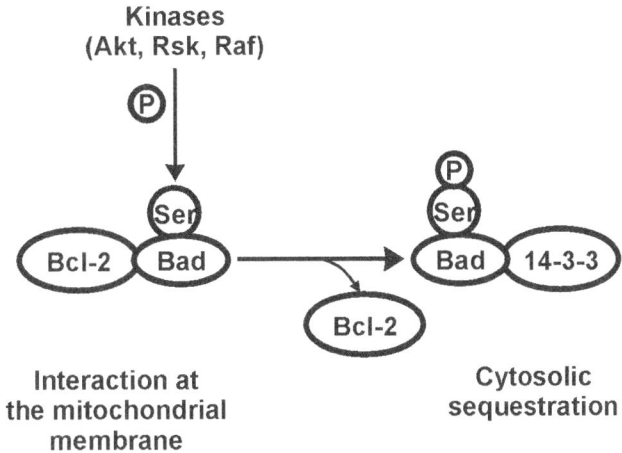

Fig. 6. **Post-translational regulation of Bad protein function in apoptosis.**

Unphosphorylated Bad heterodimerizes with Bcl-2 (or Bcl-xL) at the mitochondrial membrane and inhibits its anti-apoptotic function. Upon phosphorylation by cellular kinases (Akt, RSK, Raf), Bad associates with the 14-3-3 protein in the cytoplasm, a process associated with survival.

Interestingly, the anti-apoptotic Bag-1 protein was also shown to bind, and more significantly, activate the Raf-1 kinase (not yet studied for B-Raf or A-Raf isoforms) (Wang *et al.*, 1996), which raises the possibility that the Bag-1-Bcl-2 complex is actually responsible for targeting Raf-1 to the mitochondria (in the form of a tripartite complex), where it may phosphorylate Bad, or participate in other growth factor signal transduction processes (Fig. 7). The putative mechanisms by which Bag-1 activates Raf-1 may involve dislocation of the negative regulatory domain of Raf-1 from its catalytic domain, stabilization of Raf-1 in its active conformation once activated by other mechanisms, or protection of this kinase from inactivation by phosphatases (Morrison, 1990; Daum *et al.*, 1994; Wang *et al.*, 1996). Unlike Ras proteins, which interact with the N-terminal regulatory domain of Raf-1 (Irie *et al.*, 1994), Bag-1, similar to 14-3-3 protein, binds to the catalytic domain of Raf-1 (Wang *et al.*, 1996).

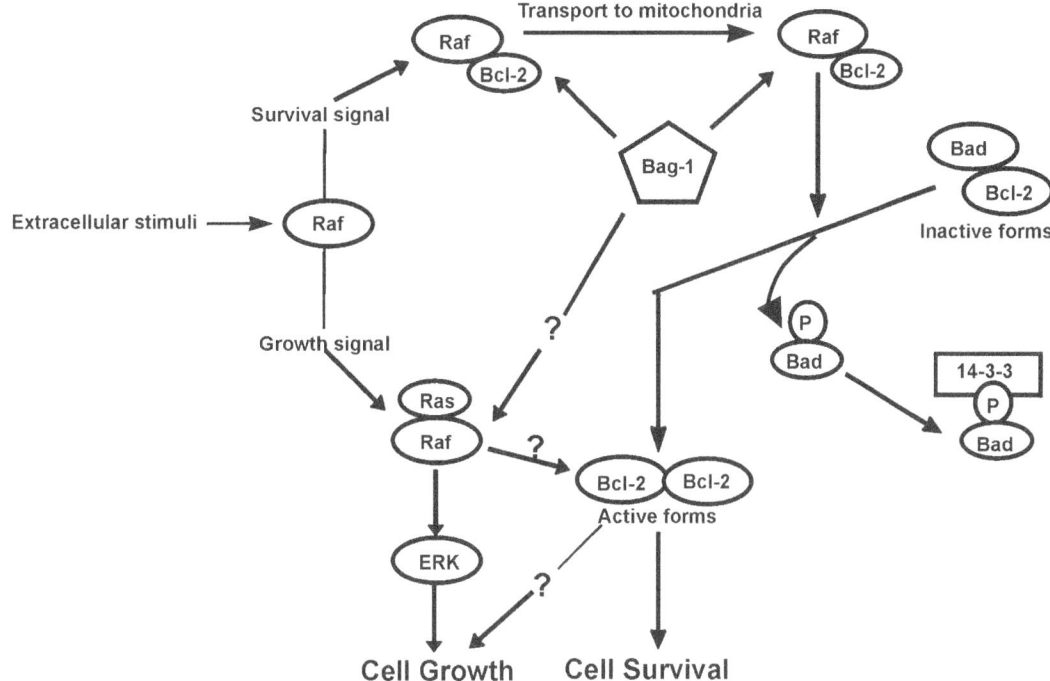

Fig. 7. **Central role of Raf-1 kinase in cell growth and survival.** Raf associates

with activated Ras leading to cell growth/proliferation by activation of the ERK cascade.

At the mitochondrial membrane, Raf phosphorylates Bad, leading to its inactivation as a

death promoter. Subcellular localization of Raf-1 as well as its kinase activity are

regulated by Bcl-2 and Bag-1.

14-3-3 proteins bind to more than 100 cellular proteins, including p53 (leading to p53 transcription factor activation and an increase in its DNA binding activity), α2-adrenergic receptors (regulating receptor localization and/or signaling), PKC (stimulating the interaction between PKC and Raf-1) and the Raf-1 or B-Raf kinases (van Hemert *et al.*, 2001). 14-3-3 proteins also mediate the rapid transition between the inactive and active forms of Raf-1 or B-Raf, but not Raf-A (Hagemann and Rapp, 1999). As such, in non-stimulated cells, 14-3-3 binds to both domains of Raf, keeping it in an inactive conformation, whereas, upon stimulation, the 14-3-3 protein is displaced from the N-terminal domain but remains attached to the catalytic (C-terminal) domain of Raf, resulting in its activation. For full activation of Raf, the presence of 14-3-3 protein is necessary (Yip-Schneider *et al.*, 2000).

One other important signaling protein involved in apoptosis is the p53 protein, which is a sequence-specific DNA- binding protein capable of transactivating genes containing p53 binding sites in either upstream regulatory regions or introns (Kastan *et al.*, 1992). p53 is a critical physiological mediator of DNA damage-induced apoptosis, but p53-independent pathways of apoptosis also exist (Murphy and Levine, 1998). In cancer, p53-dependent apoptosis controls tumor initiation, progression and response to chemotherapy (Murphy and Levine, 1998). One of the genes regulated by p53 is Bax. p53-dependent transcription of Bax leading to increased Bax protein expression results in deregulation of the ratio between pro- and anti-apoptotic proteins and formation of more Bax-Bax homodimers that induce cell death (Miyashita and Reed, 1995). On the other hand, the upstream pathway toward p53 that leads to its increased stability, transcription activation and apoptotic capacity, as part of the cellular response to stress, was found to be MEKK1/JNK (Fuchs *et al.*, 1998). The JNK-p53-Bax pathway is essential for developmental neuron death as regulated by TrkA and p75NTR (Aloyz *et al.*, 1998). Furthermore, activation of PI3-K/Akt and MEK/ERK pathways results in inhibition of pro-apoptotic p53 pathway (Aloyz *et al.*, 1998; Mazzoni *et al.*, 1999; Yamaguchi *et al.*, 2001). Bax is not necessarily the only pro-apoptotic gene targeted by p53. As such, overexpression of p53 using an adenoviral vector in cultured hippocampal neurons caused apoptosis without inducing Bax expression (Jordan *et al.*, 1997). Moreover, others reported that p53 can induce apoptosis without the transactivation of transcription (Caelles *et al.*, 1994; Wagner *et al.*, 1994). p53 induction was associated with neuronal damage induced in the CNS by kainic acid (Sakhi *et al.*, 1994), NMDA (Djebaili *et al.*, 2000), and ischemic damage (Gillardon *et al.*, 1999). Among the targets of the JNK-p53 pro-apoptotic pathway are Bax and PAG608 (p53-activated gene 608)- which is a DNA-damage-inducible gene encoding a nuclear zinc finger protein that mediates apoptosis *in vitro* (Israeli *et al.*, 1997; Varmeh-Ziaie *et al.*, 1997) and *in vivo* (Gillardon *et al.*, 1999). Furthermore, p53-inhibitors have been shown to protect neurons against death induced by ischemic or excitotoxic insults and Aβ protein (Culmsee *et al.*, 2001).

As shown in Fig. 8, there is a complex relationship between Bcl-2 family members and other components of signal transduction pathways. Caspase-8 mediated cleavage of Bid triggers Bax activation (increased expression). Bax as well as Bim may also be activated directly by various apoptotic stimuli, resulting in the formation of heterodimers with anti-apoptotic Bcl-2 proteins and their subsequent inhibition. Bcl-2 anti-apoptotic activity was inhibited when phosphorylated by JNK/SAPK in the presence of Rac (Maudrell *et al.*, 1997). However, the role of phosphorylation in the anti-apoptotic activity of Bcl-2 or Bcl-xL is unclear, since the anti-apoptotic activity of Bcl-2 was preserved when phosphorylated in response to survival factors (May *et al.*, 1994). Moreover, kinases from survival pathways, such as Rsk, Raf and Akt increase the expression of Bcl-2 and Bcl-xL by activating the transcription factor CREB (cAMP response element binding protein). By contrast, the pro-apoptotic JNK/SAPK pathway uses the transcription factor p53 to increase the expression of Bax. Caspases also interact with signaling proteins. The Ras-initiated signal transduction pathways associated with survival such as Raf/MEK/ERK and PI3-K/Akt can be turned off by proteolytic cleavage of Raf-1 and Akt and this is inhibited by caspase inhibitors or Bcl-xL (Widmann *et al.*, 1998). By contrast, neither JNK/SAPK nor p38 MAPK were cleaved (Widmann *et al.*, 1998), suggesting that this is a caspase activity specifically directed toward the inhibition of the anti-apoptotic proteins. In other instances, caspases may directly activate the pro-apoptotic signaling pathways, for example by cleavage and activation of MEKK1 (Cordone *et al.*, 1997), the upstream activator of JNK, thereby providing an apoptotic amplification loop.

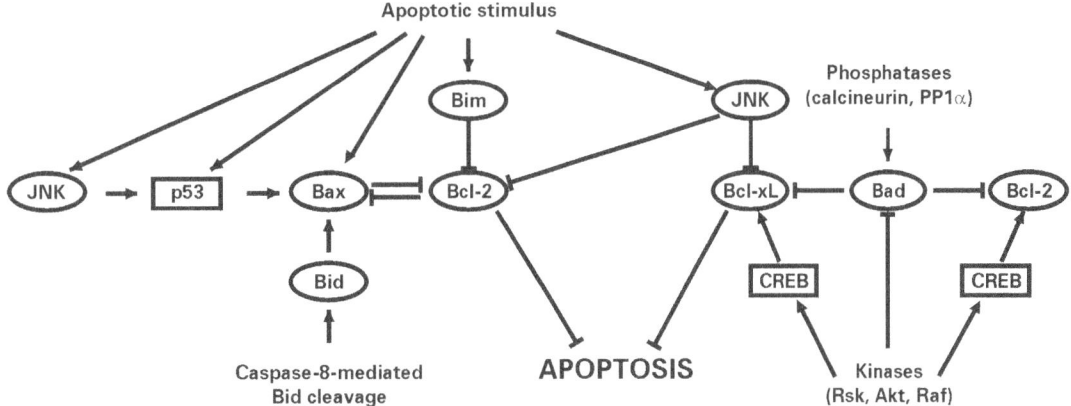

Fig. 8. **Functional relationship between signaling proteins and Bcl-2 family of proteins.** The expression of pro-apoptotic members of Bcl-2 family (Bad, Bax, Bim) is upregulated directly by apoptotic stimuli or by activation of JNK or JNK-p53 pathway. Proteins from survival pathway (RSK, Raf, Akt) upregulate the expression of anti-apoptotic proteins (Bcl-2, Bcl-xL) through a CREB-dependent mechanism.

Signaling and apoptotic pathways are thus defined by common mechanisms which involve sequential activation of protein cascades (kinases and caspases, respectively) and post-translational modifications such as cleavage and phosphorylation which are responsible for gain or loss of function. The wide variety of proteins participating in signal transduction and regulation of apoptosis suggests that there could be more common components acting on checkpoints of crucial importance in determining the cell's fate. The unraveling of the mechanisms responsible for the modulation of apoptosis, such as the specific gene expression induced by pro-apoptotic (JNK and p38) or survival (ERK and Akt) signaling pathways, will most likely contribute to the design of therapeutic strategies for diseases with an apoptotic component.

2.   Virus modulation of apoptosis

Viruses have evolved various strategies to influence the cell's commitment to apoptosis in order to increase their fitness. Viral inhibition of cellular apoptosis extends the cell life and results in maximal production and spread of progeny virions. In addition, many viruses are known to induce apoptosis (usually at the late stages of infection), a process associated with spread of progeny virions in apoptotic bodies which allows the evasion of the immune inflammatory responses and protection from host enzymes and antibodies.

Anti-apoptotic viral genes can be homologous to the cellular proteins involved in suppression of apoptosis (such as Bcl-2, IAP or FLIP) or direct inhibitors of caspases.

Moreover, anti-apoptotic viral genes may interfere with the interferon or TNFR/CD95

response, or modulate signal transduction and transcriptional regulation (Fig. 9).

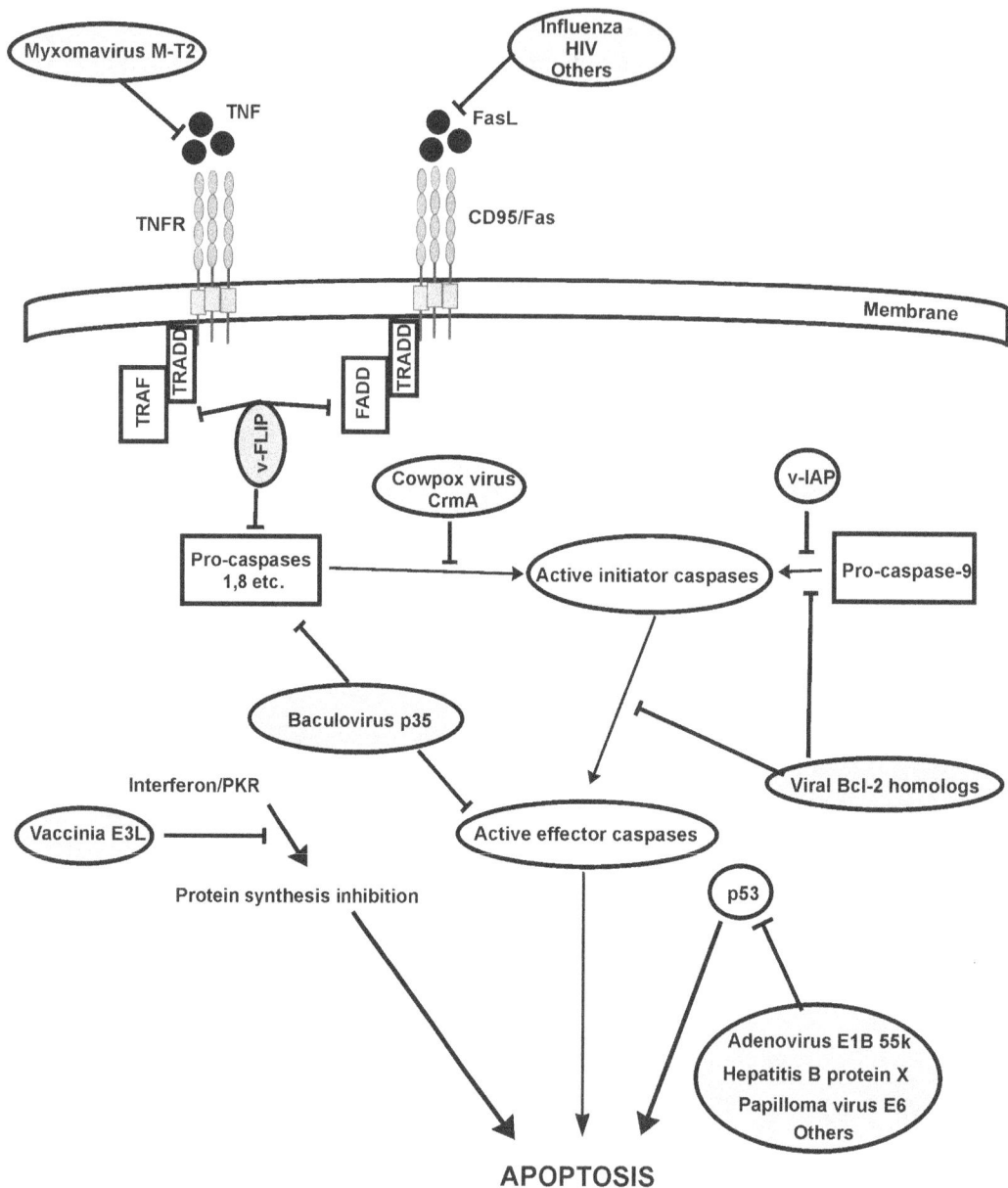

Fig. 9. **Mechanism of action of viral anti-apoptotic proteins.**

Several viruses encode homologs of the cellular Bcl-2 gene (e.g. BHRF1 gene of Epstein-Barr virus, E1B 19k of adenovirus, ORF16 of Herpesvirus Saimiri, 5-HL of African swine fever virus or KSbcl-2 of Kaposi's sarcoma associated virus) (reviewed by Hardwick *et al.*, 1998). These viral Bcl-2 homologs are able to bind and suppress the pro-apoptotic function of Bax or Bak, in the same manner as their cellular counterpart. In addition, they lack the caspase cleavage site and thereby escape the cellular regulatory mechanisms (Cheng et al., 1997; Clem *et al.*, 1998).

Baculoviruses and certain poxviruses encode homologs of cIAPs (collectively known as v-IAPs). To date, three baculovirus IAP genes (Cp-IAP, Op-IAP, and Ac-IAP) have been identified (Clem *et al.*, 1996). However, only two of these (Cp-IAP and Op-IAP) have anti-apoptotic activity. Baculovirus AcMNPV encodes the Ac-IAP gene (which is a structural homolog of c-IAP that was not yet shown to be anti-apoptotic) and the anti-apoptotic p35 gene. Baculovirus IAP proteins do not bind TRAFs like their cellular counterparts and apparently do not function as direct caspase inhibitors as they are unable to inhibit cell death induced by activated caspases-1 and 3 (Hardwick *et al.*, 1998). The mechanism of action of v-IAPs is not clear; they appear to act by binding and inhibiting pro-caspases (e.g. caspase-9) but not activated caspases (Manji *et al.*, 1997; Roy *et al.*, 1997). By contrast, the baculovirus p35 protein acts by inhibiting activated caspases (Clem *et al.*, 1996).

Poxviruses and herpes viruses encode DED-containing proteins (e.g. 71-FLIP of Herpes virus Saimiri, BORFE2 of BHV-4, E8 of EHV-2), homologous to the DED domains of caspase-8 (formerly known as FLICE). These viral proteins are collectively known as v-FLIPs (for viral FLICE-inhibitory proteins) and they are able to bind FADD and inhibit the recruitment of pro-caspase 8 to the TNF/CD95 receptor, thereby interfering with the death receptors signaling (Thome *et al.*, 1997).

A number of viral proteins act either directly or indirectly to inhibit caspase activity as a means of preventing apoptosis. Such inhibitors include the above mentioned v-IAPs, the serpin family of protease inhibitors and the unique p35 protein. Poxviruses and a murine herpes virus encode members of the serpin family (Hardwick *et al.*, 1998). Serpins are serine protease inhibitors (SPI) that regulate a wide range of cellular functions including immune and inflammatory responses. Deletion of these proteins from the serpin-containing viruses results in more attenuated viruses, in vivo, mainly because the host immune response against infection proceeds unimpaired (Upton *et al.*, 1990; Macen *et al.*, 1993). Cowpox virus CrmA (SPI-2) is an unusual serpin in that it functions as a cysteine rather than serine protease inhibitor. CrmA specifically binds and inhibits caspase-1 (involved in processing the inflammatory cytokine IL-1β) and caspases 4,5 and 8 (a subset of long prodomain caspases) by serving as a pseudosubstrate for the enzymes (Kamada *et al.*, 1997; Zhou *et al.*, 1997). CrmA inhibits cell death induced by FasL, TNFα, NGF withdrawal and extracellular matrix disruption, but does not inhibit cell death induced by DNA damaging agents (e.g. etoposide and radiation) or staurosporine. By contrast, Bcl-2 blocks apoptosis induced by radiation, DNA damage and chemical inducers but is ineffective against Fas-mediated death. This apparent dichotomy between CrmA and Bcl-2 may be due to the lack of inhibition of caspase-3 by CrmA (Nicholson *et al.*, 1995). The p35 protein of Autographica californica nuclear polyhedrosis virus

(AcMNPV) was shown to block apoptosis induced by virus infection and a wide range of other stimuli (including Fas, TNFα, radiation, etc.). p35 binds and inhibits caspase-1 and 3 by serving as a pseudosubstrate for the enzymes (similar to CrmA). However, unlike CrmA, instead of being released after its cleavage, p35 remains bound to the caspase in a stable inhibitory complex (Bump *et al.*, 1995; Xue *et al.*, 1995).

Viruses have evolved means to evade the host immune response, by inhibiting the antiviral, apoptotic effects of interferon. For example, vaccinia virus protein E3L target cellular anti-viral mechanisms involving activation of ds RNA-activated protein kinase (PKR). PKR is activated by double-stranded (ds) RNA intermediates that occur during virus replication or by an interferon-dependent mechanism. It phosphorylates the translation initiation factor eIF-2α followed by subsequent shut off of cellular protein synthesis. The activated PKR-induced apoptosis is efficiently inhibited by the E3L protein by direct inhibition of the enzyme (Kibler *et al.*, 1997).

Viruses may also inhibit apoptosis induced by Fas and TNFα (through activation of their receptors, Fas/CD95 and TNFR, respectively), thereby interfering with the host cell defense mechanisms exerted by T-cells and macrophages. Apart from interfering with the apoptotic signal transduction (as described for v-FLICEs and v-IAPs), viruses have also evolved other means to interfere with this pathway. For example, Myxoma virus (a poxvirus that causes an immunosuppressive disease called myxomatosis in rabbits) encodes the M-T2 protein that binds to and inhibits the action of TNFα (Schreiber *et al.*, 1997). Additionally, cowpox produces two secreted proteins CrmB and CrmC, that resemble the extracellular domains of TNFR and compete effectively for TNF ligand binding (Roulston *et al.*, 1999).

Viral anti-apoptotic strategies that involve modulation of other signal transduction pathways and regulation of transcription, also exist. For example, polyomavirus middle T antigen and the hepatitis B virus X protein directly activate the pro-survival PI3-K/Akt pathway which leads to Bad phosphorylation and inhibition of its pro-apoptotic function (Dahl *et al.*, 1998; Webster *et al.*, 1998; Lee *et al.*, 2001). Also, activation of the PKC pathway was suggested to be involved in the protective activity of SV40 large T antigen in response to Fas-dependent apoptosis (Rouquet *et al.*, 1995). The Epstein-Barr virus (EBV) anti-apoptotic protein LMP-1 targets TRAF family members (thereby interferes with TNF receptor signaling). It also induces increased expression of anti-apoptotic proteins, such as Bcl-2 (Henderson *et al.*, 1991), A20 (Laherty *et al.*, 1992), and Mcl-1 (Wang *et al.*, 1996), and inhibition of c-Myc synthesis (Kawanishi, 1997). These pluripotent activities of LMP-1 are believed to contribute to establishment of EBV latency. Another protein of EBV, the Bcl-2 homolog BHRF1 was reported to bind directly to R-Ras (like Bcl-2). This finding led to the hypothesis (not yet studied) that BHRF1 may contribute to activation of Ras-dependent signaling pathways leading to cell survival (Theodorakis *et al.*, 1996).

In addition to interfering with existing apoptotic or signaling pathways, viruses have the ability to modulate apoptosis-related gene transcription. Virus encoded transcription regulators protect against cell death by up- or downregulating genes involved in apoptosis. For many viruses, replication depends on the induction of the S phase of cell cycle, which often leads to increased levels of the transcription factor p53.

Activated p53 leads to apoptosis and, therefore, these viruses have evolved specific means of counteracting p53 to maximize progeny production. Several viral anti-apoptotic proteins (Teodoro and Branton, 1997; Hardwick *et al.*, 1998) block p53-dependent apoptosis either by repression of its transactivation activity (Ad E1B 19 K, E4orf6, HBV pX), sequester it in the cytoplasm (Ad E1B 55 K) or targeting it to degradation via the ubiquitin pathway (HPV E6). The ability of viruses to interfere with p53-dependent apoptosis may contribute to neoplastic transformation associated with viral pathogenesis.

Viruses can also cause cell death. Virus-induced apoptosis represents a very effective strategy for releasing progeny virions in membrane bound apoptotic bodies without stimulating an inflammatory response, thereby undermining the natural host defense. This mechanism is particularly important for the release of non-enveloped, non-lytic viruses (Hardwick *et al.*, 1998). Virus-induced apoptosis also contributes to disease pathogenesis. In this case the cellular mechanisms are essential in determining the cell's fate. As described for the Sindbis virus, successful inhibition of apoptosis will shift the lytic cycle to a persistent non-lytic infection which may be preferable to the loss of neurons (Hardwick *et al.*, 1998). Krakauer and Payne (1997) developed a mathematical model to account for the two opposite ways by which viruses influence the host cell's fate. According to this model, the factors which determine whether the viruses inhibit or stimulate apoptosis are: rate of cell death, rate of virus release and efficiency of virus production during the lytic cycle. The model suggests that viruses evolve towards increasing their reproductive fitness, being more cytopathic in cells with increased lifespan and low efficiency of virus extrusion, and less cytopathic in cells with low lifespan and high efficiency of virus release.

Viruses known to induce apoptosis are herpes viruses, baculoviruses, adenoviruses, alphaviruses (e.g. Sindbis), lentiviruses (e.g. HIV-1), influenza, parvoviruses (e.g. B19), reovirus and chicken anemia virus. The mechanisms by which they induce apoptosis include p53-dependent and independent pathways or upregulation of Fas ligand. In addition, viruses may alter the expression of anti-apoptotic proteins, such as Bag-1. For example, Coxsackievirus B3-induced apoptosis in myocytes is associated with downregulation of Bag-1 expression, a mechanism that may contribute to virus-induced myocarditis or dilated cardiomyopathy (Peng *et al.*, 2001). However, for most viruses the exact genes and the apoptotic pathways that they stimulate are not yet known. In some cases the viral anti-apoptotic proteins function as apoptosis brakes: when the brake is released (the anti-apoptotic proteins are not present) the virus activates its pro-apoptotic proteins (e.g. baculovirus p35 and adenovirus E1A) which trigger cell death. The mechanisms by which a pro-apoptotic viral protein exerts its function are often cell specific. For example, the adenovirus E1A protein induces apoptosis in a p53-dependent or independent manner, depending on the cell type (Hardwick *et al.*, 1998). In certain cell types, E1A induced apoptosis requires the presence of another adenovirus protein E4orf4 which binds protein phosphatase 2A (PP2A), thereby potentially affecting the phosphorylation status of proteins involved in signaling and apoptotic pathways (Marcellus *et al.*, 1996).

In conclusion, a variety of viruses encode genes which modulate cell death which suggests a role for apoptosis in the viral life cycle. Such a role may involve induction of apoptosis as a means for the release of non-enveloped, non-lytic viruses and inhibition of apoptosis as a mechanism for establishing latency or inhibiting the immune response.

C.   Herpes Simplex Viruses

Herpes simplex viruses type 1 (HSV-1) and type 2 (HSV-2) are large (at least 152 kb) DNA viruses (reviewed by Aurelian, 2000). The variability in size is mainly due to the variation in the number of reiterations of specific terminal and internal sequences. HSV DNA consists of two covalently linked components, designated as long (L) and short (S). Each component consists of unique sequences (U_L or U_S, respectively) flanked by relatively large inverted repeats. The inverted repeats sequences flanking U_L are *ab* and *b'a'*, whereas those flanking U_S are *a'c'* and *ca*. HSV encodes at least 84 different proteins, out of which at least 45 genes are dispensable for viral growth. These genes include those whose products enable the virus to attach and enter cells, to fine-tune expression of individual genes and to block cellular and immune responses to viral infection. It should be noted that these genes are not truly dispensable: viruses lacking these genes have not been isolated from human carriers, and viruses from which these genes have been deleted by genetic engineering frequently exhibit reduced capacity to multiply and spread in experimental animals. The fact that so many genes are dispensable is the basis of HSV gene therapy which employs HSV recombinants that have foreign therapeutic genes inserted into their genomes. Out of the 84 genes, fourteen map in the U_S, four map in each of the sequences flanking U_L, one maps in each of the sequences flanking U_S, and the remainder map in U_L. The genes mapping in the repeats are present

in two copies per genome. The virion (150 to 200 nm in diameter) consists of four

components: 1) an electron-dense core, that contains the linear, double-stranded DNA; 2)

an icosadeltahedral capsid; 3) an amorphous layer of proteins, called tegument, that

surround the capsid; and 4) an envelope (Fig. 10).

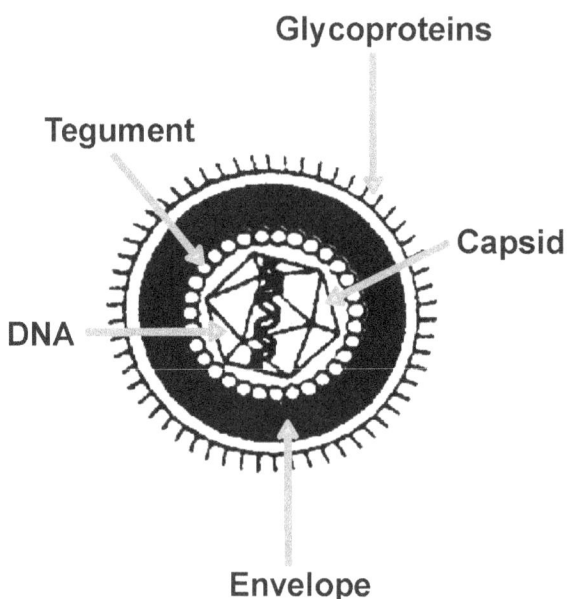

Fig. 10.   **Schematic diagram of HSV virion.**

Approximately half of the structural proteins are in the tegument, and some are involved in initiation of the replicative cycle. They include UL41 (also known as virion host shut off or vhs), which is responsible for decreased host mRNA stability and inhibition of host cell translation; UL48 (also known as VP16), which is responsible for the transactivation of immediate-early (IE) genes; UL13, which has protein kinase (PK) activity; RR1 (also known as ICP6 or ICP10, for HSV-1 or HSV-2, respectively), which also has PK activity and is required for optimal expression of HSV-2 IE genes (Smith and Aurelian, 1997); and IE proteins ICP4 and ICP0, which regulate the expression of other viral genes (Yao and Courtney, 1991).

To initiate infection, HSV must attach to cell-surface receptors, fuse its envelope to the plasma membrane, and allow the de-enveloped capsid to be transported to the nuclear pores, where the DNA is released into the nucleus. The main events in virus replication that occur in the nucleus include transcription, DNA synthesis, capsid assembly, DNA packaging, and envelopment (Fig. 11).

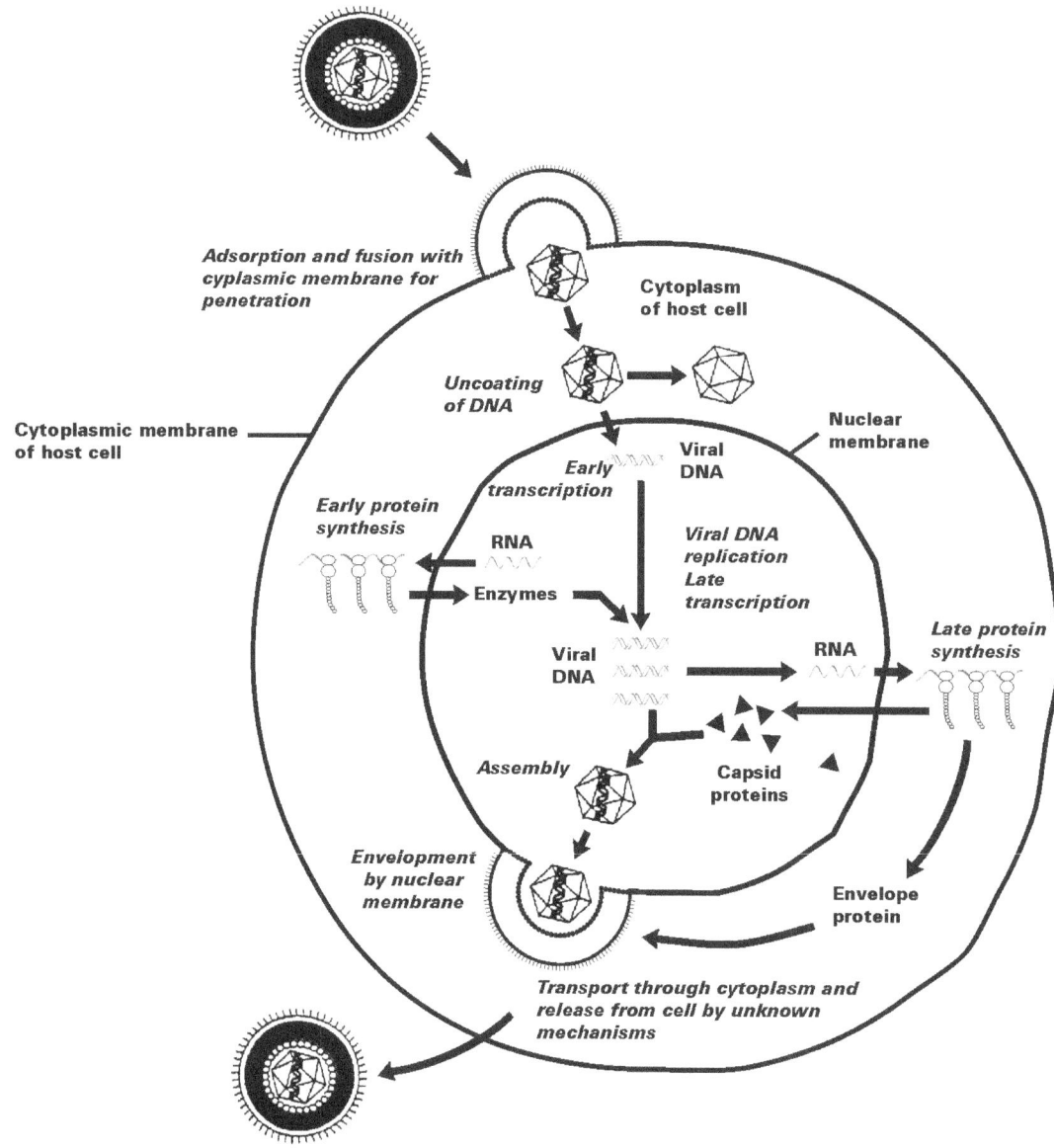

Fig. 11   **Schematic representation of HSV replication in susceptible cells.**

The viral envelope contains surface glycoproteins (designated gB, gC, gD, gE, gG, gH, gI, gJ, gK, gL, and gM). The initial adsorption of HSV to glycosaminoglycans on the cell surface is mediated by gC and/or gB (Herold *et al.*, 1991, Herold *et al.*, 1994). This is followed by a sequence of interaction of gD with cellular receptors and pH-independent fusion of the viral envelope with the cell membrane which is mediated by gB, gD, and the complex of gH and gL (Nicola *et al.*, 1998). The cellular receptors for HSV were designated herpes virus entry mediators (HVEMs) and the first discovered (HveA) belong to the TNF receptor family. It was established that HveA interacts with several members of the TNFR-associated factor (TRAF) family and transduces a signal via activation of JNK and Jun-containing-AP-1 transcription factor (Marsters *et al.*, 1997). Additional receptors include poliovirus receptor-related proteins 1 and 2 (PRR1 and PRR2, respectively) which are also known as HveC and HveC, respectively, and the herpes immunoglobulin-like receptor (HIgR), which is a splice variant of PRR1 (Aurelian, 2000).

The synthesis of viral gene products takes place in three steps. The first group of viral gene products consists of six proteins known as IE (or α) proteins which reach maximal expression around 2-4 hrs post-infection (p.i.) but continue to accumulate until late in infection. Of these, four (ICP0, ICP4, ICP22, ICP27 and RR1) regulate the replicative cycle of the virus, and one (ICP47) blocks the presentation of antigenic peptides on the infected cell surface (Aurelian, 2000). IE proteins share common features: 1) the presence of a TAATGARAT sequence in their promoter; 2) expression in the absence of other viral protein synthesis, and 3) activation by VP16, which is brought into the cell by the virion. VP16 forms a multicellular DNA binding complex (on the TAATGARAT motif) with the cellular proteins oct-1 and HCF (which is required for progression through G1 phase of cell cycle) triggering the transcription of IE genes (Roizman and Sears, 1996). IE proteins are essential for the synthesis of a second set of viral gene products, known as early (E, or β) proteins which reach a peak at 5-7 hrs p.i.. Most of the E proteins are responsible for viral nucleic acid metabolism and are the main targets of antiviral chemotherapy (i.e., viral thymidine kinase and viral DNA polymerase). The late (L, or γ) class of proteins is for the most part of the structural components of the virion. They assemble to form the capsid and tegument, and they incorporate into nuclear membranes for eventual envelopment of virions.

In the process of virus assembly, a procapsid is formed from scaffolding and capsid proteins. The newly synthesized DNA is cleaved into unit length molecules and packaged into virions. Capsids containing DNA attach to the nuclear surface of the inner nuclear membrane and are rapidly enveloped and released into the space between the inner and outer nuclear membranes. Here, the virions become encased in transport vesicles and are transported through vesicles formed by fragmented and dispersed Golgi stacks to the extracellular space (a process that takes about 18 hrs) (Whitley et al., 1998).

The productive HSV infection of susceptible (or permissive) cells results in cell lysis. The host cell metabolism is shut off early in infection. Thus, host DNA synthesis is shut off, host protein synthesis declines very rapidly, and glycosylation of host protein ceases. This shut off is faster and more effective in HSV-2- than in HSV-1infected cells (Roizman and Sears, 1996). The first stage is mediated by the vhs protein (which is brought into the cell packaged into the virion tegument). Besides destabilization and degradation of host mRNA (thereby removing the preexisting host mRNA from the pool of translatable messages), vhs is also responsible for destabilization and degradation of viral IE, E and L mRNAs, allowing for an efficient regulation of temporal kinetics of IE to E, and E to L transitions. The second stage requires *de novo* synthesis of viral proteins and appears to be dependent on the presence of IE gene ICP27 (Hardwicke and Sandri-Goldin, 1994). Other proteins, such as the UL13 PK and VP16 (which forms a complex with vhs) seem to be involved in vhs activity (Overton *et al.*, 1994; Lam *et al.*, 1996).

HSV-1 and HSV-2 have an overall DNA homology of 50 % (which is higher or lower in some genomic regions) (Aurelian, 1998). They have a predilection for different body sites (HSV-1 is mostly associated with oral lesions, whereas HSV-2 causes genital lesions) and are, therefore, generally transmitted by different routes. Yet, both involve close personal contact (virus must come in contact with mucosal surfaces or abraded skin for infection to be initiated). Animal vectors of human HSV infections have not been described, and humans remain the sole reservoir for transmission to other humans. Susceptible individuals (namely those without preexisting antibodies to HSV) develop primary infection after their first exposure to either HSV-1 or HSV-2. The histopathological characteristics of HSV infection reflect virus-mediated cellular lysis and associated inflammatory responses. Viral infection induces ballooning of susceptible cells, with condensation of chromatin and degeneration of cellular nuclei, generally within the parabasal and intermediate cells of the epithelium. Infected cells lose intact plasma membranes and form multinucleated giant cells (polykaryocytosis is the cell alteration used in the cytologic diagnosis of HSV infections). When cell lysis occurs, a clear (vesicular) fluid containing large quantities of virus appears between epidermal and dermal layers. The vesicular fluid contains cellular debris, inflammatory cells and, often, multinucleated giant cells. In the dermis, there is an intense inflammatory response. When healing occurs, the vesicular fluid becomes pustular, with the recruitment of inflammatory cells and subsequent formation of scabs. When mucous membranes are involved, the vesicles are replaced by shallow ulcers. The spectrum of disease caused by HSV includes infection of the mucous membranes (gingivostomatitis, herpes labialis and genital herpes infections), keratoconjuctivitis, neonatal HSV infection, visceral HSV infections in immunocompromised hosts, Kaposi's varicella-like eruption, and erythema multiforme (Whitley *et al.*, 1998).

1. HSV infection of the nervous system

Upon infection of mucosal membranes or abraded skin, both HSV-1 and HSV-2 are transported by axonal retrograde movement involving microtubules to the sensory ganglia (trigeminal and sacral for HSV-1 and HSV-2, respectively), where they establish latency (Fig. 12). Latency refers to the maintenance of the viral genome in a largely non-transcribed episomal state providing a virus reservoir for the entire life of the infected individual by means of periodic virus reactivation whereby virus is transported by reverse axonal transport back (or close) to the dermatome of entry (recurrent infection). Reactivation occurs following a variety of local or systemic stimuli such as physical or emotional stress, fever, exposure to UV light, tissue damage and immunosuppression. Recurrences occur in the presence of both humoral and cell-mediated immunity.

Extensive studies of ganglia harboring latent HSV did not reveal any viral proteins expressed. However, a viral RNA transcript designated LAT (latency associated transcript) has been detected in latently-infected neurons, largely confined to the nucleus. LAT is now believed to be a stable intron, approximately 2 kb in length, derived from a primary transcript of approximately 8 kb. Despite its presence, LAT is not absolutely required for establishment of latency and/or reactivation (Roizman and Sears, 1996).

It is still an unresolved issue whether latency establishment is preceded or not by viral replication since studies confirming both possibilities have been reported (Speck and Simmons, 1991; Margolis *et al.*, 1992; Jacobson *et al.*, 1998). There is evidence that infection of sensory neurons associated with latency establishment and viral reactivation does not involve the death of neurons-which will result from lytic virus infection (Steiner and Kennedy, 1991; Whitley *et al.*, 1998). Thus, neurons are non-permissive at the time they harbor virus in the latent state, and in the ganglia permissivity is transient. When placed in culture, however, neurons taken from a latently-infected tissue become permissive. In these conditions, viral replication ensues. Moreover, it was reported that NGF is essential for maintenance of virus in the latent state, and NGF withdrawal triggers virus reactivation (Wilcox and Johnson, 1987). The mechanisms of latency establishment and/or reactivation are still unclear. It was suggested that specific cellular factors in sensory neurons would inhibit virus replication and thereby the expression of viral genes required for virus replication. Kristie *et al.* (1999) showed that HCF (host cell factor, which is essential for the transactivation of the viral regulatory IE genes), is sequestered in the cytoplasm of infected sensory neurons, while it is localized in the nucleus under conditions that support virus reactivation, which suggests its involvement (presumably a switch role) in the lytic/latent processes associated with HSV infection.

Both HSV serotypes are able to reach and infect the CNS (Fig. 12). HSV infection of the CNS may occur in the newborn and adults (causing aseptic meningitis or HSE).

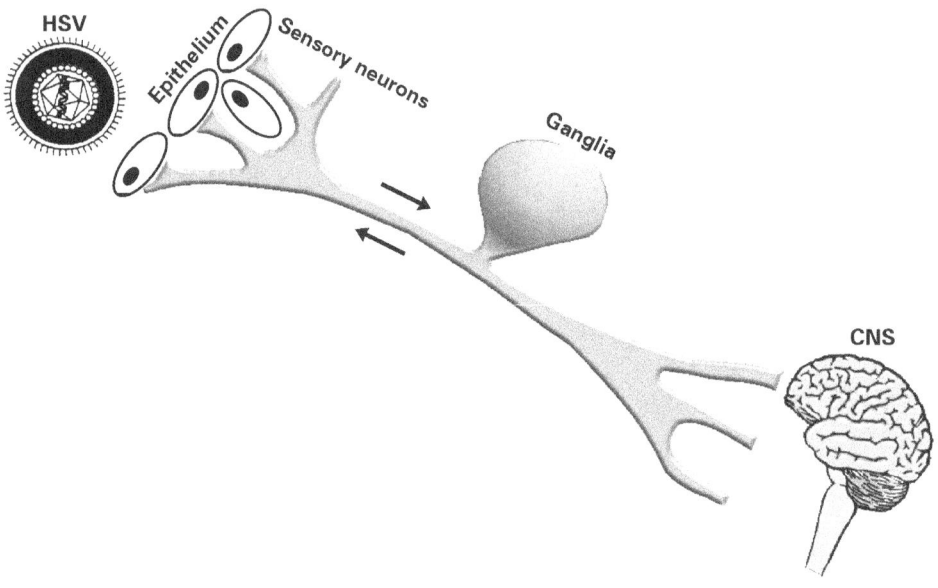

Fig. 12.   **Schematic diagram of HSV life cycle.** Upon infection of mucosal membranes or abraded skin, HSV travels by retrograde axonal transport to sensory ganglia where it establishes latency. In certain conditions, not yet elucidated, HSV reaches the CNS, most likely via a neuronal route.

Neonatal infection of the CNS occurs in approximately 50 % of infants who have HSV-2 infection during the neonatal period, and it is associated with 40-50 % mortality even when treated. The route of HSV-2 to the brain in infants that have multiorgan disseminated infection is likely the blood and is associated with a diffuse encephalitic process that results in generalized encephalomalacia. When disease involves only the CNS and is not disseminated, neuronal transmission of virus to the CNS tends to result in initial unitemporal involvement, with subsequent bitemporal disease as illness progresses. In adults, HSV-2 can cause aseptic meningitis, predominantly among patients who have primary genital disease. The outcome in these cases is excellent, even without antiviral therapy.

HSE is the most common cause of non-epidemic, sporadic, acute focal encephalitis in the United States and it is due to HSV-1. HSV-2, which is the most common cause of genital herpes and, not infrequently, viral meningitis, results in encephalitis only very rarely, specifically in immunocompromised hosts. HSE prevalence is approximately 1 in
250,000 to 500,000 individuals annually. The mortality of HSE (in the presence of antiviral therapy) is of 50-60 % and survivors are left with neurological sequellae involving impairments in memory, cognition and personality (Whitley and Kimberlin, 1999).

HSE predominantly involves localized regions of the brain comprising the limbic system. Thus, viral antigens in HSE brains are concentrated mainly in the medial and inferior temporal lobes, hippocampus, amygdaloid nuclei, olfactory cortex, insula and cingulate gyrus (Damasio and Van Hoesen, 1985). An explanation for HSE localization was sought in terms of the route by which the virus may reach the brain. In particular, two pathways have been suggested, involving intraneuronal spread of virus along the olfactory nerves and tracts, or along branches of the trigeminal nerve innervating the basal meninges. Access of virus via the olfactory route would better explain the localization of the lesions since the limbic system includes the central olfactory pathways (Esiri, 1982). Indeed, intranasal infection of mice with HSV-1 demonstrated the spread of virus to the brain via the olfactory pathways (Tomlinson and Esiri, 1983). The difference in neurological outcome of CNS infection with HSV-1 and HSV-2 was never investigated. To allow discrimination between HSV-2 and HSV-1 infection of the brain, a very useful model is the rat, because it more closely resembles the human HSE pathology which includes HSV-1 targeting and destruction of hippocampal neurons (Bergstrom *et al.*, 1991). Bergstrom *et al.* (1991) showed that upon intracranial infection of rats, HSV-1 caused neuronal destruction whereas cells infected with HSV-2 were protected. However, the mechanism involved in neuronal death was not investigated. The mechanisms responsible for the different neurological outcome following infection with the two viruses may potentially involve serotype-specific anti-apoptotic proteins. Identification of viral gene(s) with anti-apoptotic activity in hippocampal neurons could lead to the development of novel therapeutic strategies for neurodegenerative disorders, such as AD.

2.   HSV effect on cellular signaling

Viruses activate signal transduction pathways to maximize their replication/progeny production and use them to leave a long lasting mark on the cell by inducing cell transformation or modulating apoptosis. They have specific strategies for activating signal transduction and taking control of the host cell machinery. Ras-initiated signaling pathways are main targets for viral takeover and modification. For example, vaccinia virus encodes a protein that mimics epidermal growth factor (EGF) in terms of its ability to stimulate cognate receptors (King *et al.*, 1986). A Herpesvirus Saimiri protein (STP-C488) associates with Ras and activates it (Jung and Desnoiriers, 1995) and a coxsackievirus protein (Sam68) binds Ras-GAP and inactivates it, thereby activating Ras (Huber *et al.*, 1999). Src tyrosine kinases are activated by protein X of Hepatitis B virus (HBx) resulting in activation of Ras/Raf/ERK pathway and stimulation of transcription (Klein *et al.*, 1997; Klein *et al.*, 1999). As major players in the mitogenic/proliferative cascade, ERKs are targeted by a wide variety of viruses. HIV incorporates ERK in its virions as a requirement for the translocation of reverse transcriptase to the nucleus (Jacque *et al.*, 1998). SV40 small T antigen binds protein phosphatase 2A, thereby preventing it from dephosphorylating MEK and ERK2 and prolonging their activation state (Sontag *et al.*, 1993). Also, the middle T antigen of polyomavirus mediates cellular transformation by associating with the adaptor protein Shc therefore activating Ras and the ERK pathway through Grb2 and Sos (Robinson and

Cobb, 1997).

HSV-1 and HSV-2 were originally considered unconventional DNA transforming viruses because cell transformation could be achieved with DNA sequences which do not specify viral proteins (Reyes *et al.*, 1979). However, HSV-2 transforming function was subsequently localized to a DNA fragment which codes for a novel Ser-Thr protein kinase (PK) located at the amino terminal domain of the large subunit of ribonucleotide reductase (R1, also known as ICP10) (Jariwalla *et al.*, 1980; Hayashi *et al.*, 1985). ICP10 PK belongs to a subfamily of growth factor receptor Ser-Thr PKs that are involved in the regulation of diverse cellular processes including proliferation, differentiation and apoptosis (Smith *et al.*, 1991; Hunter *et al.*, 1995; Tian *et al.*, 1995). Studies of human cells transformed by HSV-2 R1 PK indicated that it functions as an activated growth factor receptor that stimulates the Ras/MEK/ERK pathway (Smith *et al.*, 1994) by binding Grb2-Sos complex as well as inactivating Ras-GAP. The individual contribution of each one of these two functions to pathway activation is still unclear. However, there is considerable evidence that activation of Ras/MEK/ERK mitogenic pathway by the PK activity of R1 as well as the R1 PK itself are required for maintenance of the transformed phenotype in non-neuronal cells (Aurelian, 1998). By contrast HSV-1 hijacks the host non-neuronal cell machinery by activating JNK/SAPK and p38 MAPK (Zachos *et al.*, 1999) and its R1 PK does not have transforming activity (Aurelian and Smith, 2000). Activation of these stress-activated pathways leads to enhanced HSV-1 replication. Notwithstanding, activation of JNK/SAPK and p38 MAPK in HSV-1 infected cells may be part of a cellular response to viral infection, namely commitment to apoptosis.

Maneuvering the signal transduction pathways is also an important mechanism of evading the host defense. Cells respond to viral infections, in part by downregulating the overall rate of protein synthesis. This translational control occurs largely through the modification of the translation initiation factor eIF2 (Dever, 1999). Phosphorylation of eIF2 on Ser 51 (on its subunit α) impedes eIF2 function, resulting in halting translation initiation and shutdown of global cellular protein synthesis. One of the kinases able to phosphorylate eIF2α is PKR (the host IFN-induced protein kinase), which is activated by viral transcripts in the form of ds RNA produced during viral replication. As a result of PKR phosphorylation of eIF2α, the cellular translational machinery is incapacitated and viral protein synthesis and replication are restricted within the infected cell. Not to be outdone, viruses have evolved means to counteract PKR-mediated translational block, including inhibition of PKR activation (by blocking its catalytic site or interfering with its achieving the active conformation) or by directly regulating the eIF2α phosphorylation. One of the HSV genes that directly regulates eIF2α phosphorylation is ICP34.5 (also known as $\gamma_1 34.5$), which binds and activates the host protein phosphatase1α (responsible for dephosphorylating eIF2α) (Tan and Katze, 2000), thereby circumventing the translational block resulting from PKR activation (Tan and Katze, 2000). Cellular homologs of the ICP34.5 viral gene include Myd116 (expressed in myeloid leukemia cells induced to differentiate by IL-6) and GADD34 (a protein induced by growth arrest and DNA damage) (Roizman and Sears, 1996). Since PKR was also involved in apoptosis, a possible role for ICP34.5 in modulation of cell death induced by shut off of protein synthesis was suggested (Chou and Roizman, 1992).

In an effort to clarify the latency establishment/reactivation mechanisms associated with HSV infection, and/or due to the growing interest in using HSV-based vectors in gene therapy of the brain (Fink and Glorioso, 1997), there is an increasing number of reports dealing with modulation of signaling proteins by HSV, especially in rat pheochromocytoma PC12 cells (which differentiate into cells resembling sympathetic neurons after treatment with NGF). Significantly, in this system, latently (or quiescently) infected neurons reactivate the virus upon NGF withdrawal (Block *et al.*, 1994).

The major latency-associated transcript (LAT) is expressed in a stable, non-linear form in PC12 cells productively infected with HSV-1 (Rodahl and Haarr, 1997). Moreover, LAT expression is significantly enhanced by NGF and sodium butyrate (either alone or in a synergistic effect) in neuronally differentiated PC12 cells. Activation of LAT promoter by these two factors requires Ras activation and the activation of Raf is sufficient to induce LAT expression (Frazier *et al.*, 1996). Significantly, treatment of PC12 cells with EGF (which triggers Ras/Raf activation and c-fos induction leading to a mitogenic response in these cells) does not activate the LAT promoter, suggesting that it responds to the specific induction of Ras/Raf pathway by neurotrophins (Frazier *et al.*, 1996). Thus, neurotrophins, such as NGF, which are required for survival and maintenance of sensory neurons, may also serve to prevent virus reactivation through induction of LAT.

Other factors, such as glucocorticoids and heat stress are also known to reactivate HSV from latently infected neurons. In neuronally differentiated PC12 cells, these factors act to stimulate viral replication (Hardwicke and Schaffer, 1997; Danaher *et al.*, 1999).

During the response of infected PC12 cells to heat shock, the IE protein ICP4 and host HSP72 were expressed at an early time (3 hrs of treatment or 0 hr recovery time), suggesting a role for ICP4 in coordinating reactivation from latency (Danaher *et al.*, 1999). ICP4 is a DNA-binding protein required for virus growth and acts both as a negative regulator of IE gene expression and as a positive transactivator of E and L proteins (Roizman and Sears, 1996). The interaction and post-translational modification of ICP4 by protein kinase A (PKA, known to associate with the transcriptional machinery and modify the activities of proteins that bind to DNA) seem to be required for optimal virus growth, most likely by promoting further protein interactions (Xia *et al.*, 1996).

In conclusion, HSV has evolved means to interact with signaling proteins that can be thus diverted to perform novel functions. These viral activities are involved in enhancing virus growth, establishing latency/reactivation or evading the host immune response, and may also have a significant role in modulation of apoptosis.

**3.	The role of ICP10 PK in the virus life cycle**

Ribonucleotide reductase (RR) plays an essential role in DNA synthesis of eukaryotic and prokaryotic cells. Similar to mammalian and bacterial enzymes, the HSV RR consists of two heterologous subunits of 140 kDa (RR1) and 38 kDa (RR2). The large subunit (RR1) is designated ICP6 for HSV-1 and ICP10 for HSV-2, and the respective gene is encoded within the U_L region of the viral genome. The RR gene is not essential for virus growth in dividing cells but it is required for virus replication in serum-starved or neuronal cells (Jacobson *et al.*, 1989). RR2 is an E gene (its synthesis requires a functional ICP4). By contrast, RR1 is an IE gene, because : 1) it is synthesized at early times (2 hrs p.i.) and it continues to accumulate throughout the replicative cycle; 2) its synthesis occurs in the absence of other viral proteins, does not require a functional ICP4 but is increased by ICP0; 3) it has the TAATGARAT sequence on its promoter, which leads to its activation by the VP16/oct-1 complex. Significantly, the RR1 promoter has response elements for AP-1 transcription factors (not present in other viral genes) which are required for basal expression of HSV-2 RR1 (Aurelian, 1998). Proteolytic degradation studies have associated RR activity with the carboxyl two thirds of the HSV RR1 molecule, leading to the conclusion that the amino-terminal domain is functionally distinct (Ingemarson and Lankinen, 1987). This N-terminal domain shows only 38 % homology between HSV-1 and HSV-2, as compared to 93 % for the carboxyl two thirds of the molecule (Nikas *et al.*, 1986). Phylogenetic studies indicated that the N-terminal domain of HSV-2 RR1 codes for a PK originated from a Ser-Thr kinase growth factor receptor (Aurelian, 1998). Indeed, functional studies from our laboratory (Chung *et al.*, 1989; Luo and Aurelian, 1992; Smith *et al.*, 1994; Hunter *et al.*, 1995; Smith et al., 2000) indicated that the N-terminal domain (amino acids 1-447) of the HSV-2 RR1 (ICP10) protein has intrinsic Ser-Thr kinase activity. The catalytic domain of ICP10 PK contains 8 conserved catalytic motifs and is preceded by a single transmembrane (TM) helical segment followed by a basic amino acid (responsible for TM anchorage within the cell membrane), a short extracellular (EC) domain and a signal peptide (SP)-common to membrane-associated proteins (Perlman and Halverson, 1983; Chung *et al.*, 1989) (Fig.

13). The minimal PK domain (amino acids 1 to 283) includes amino acids 176, 209 and

259, and contains the PK catalytic core. The two proline-rich consistent with SH3-

binding motifs and an SH2-domain were also identified (Smith *et* al., 1994). ICP10 is

associated with the plasma membrane and deletion of the TM domain eliminates cell-

surface localization and its PK activity (Smith *et al.*, 1994; Hunter *et al.*, 1995). Amino

acids 396 to 405 (and also to a lower extent, 149 to159) are required for binding of Grb2-

Sos, whereas amino acids 112 to 122 and 136 to 146 (which include pThr117 and

pThr141) are required for binding Ras-GAP SH2 domain. Known substrates for ICP10

PK include calmodulin, some but not all histones, Ras-GAP and the RR2 viral protein

(Aurelian, 1998).

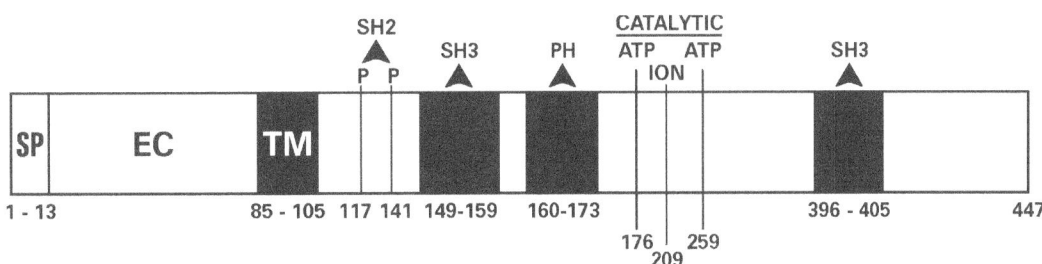

Fig.13.   **Schematic diagram of ICP10 PK structure.**

ICP10 is involved in modulation of signal transduction pathways (Fig. 14). It binds the Grb2-Sos complex, thereby bringing the complex in the vicinity of Ras leading to its activation. To overcome the downregulatory activity of RasGAP, the activated ICP10 also binds Ras-GAP and phosphorylates/ inactivates it. This dual action of ICP10 leads to increased levels of activated Ras. This in turn stimulates the downstream kinase cascade (Raf/MEK/ERK) which in non-neuronal cells culminates in increased expression of AP-1 transcription factors (such as c-fos) and mitogenesis (Aurelian, 1998; Smith *et al.*, 2000). This pathway activation provides an amplification loop that leads to increased ICP10 expression and, moreover, it is required for timely onset of virus growth (Smith *et al.*, 2000).

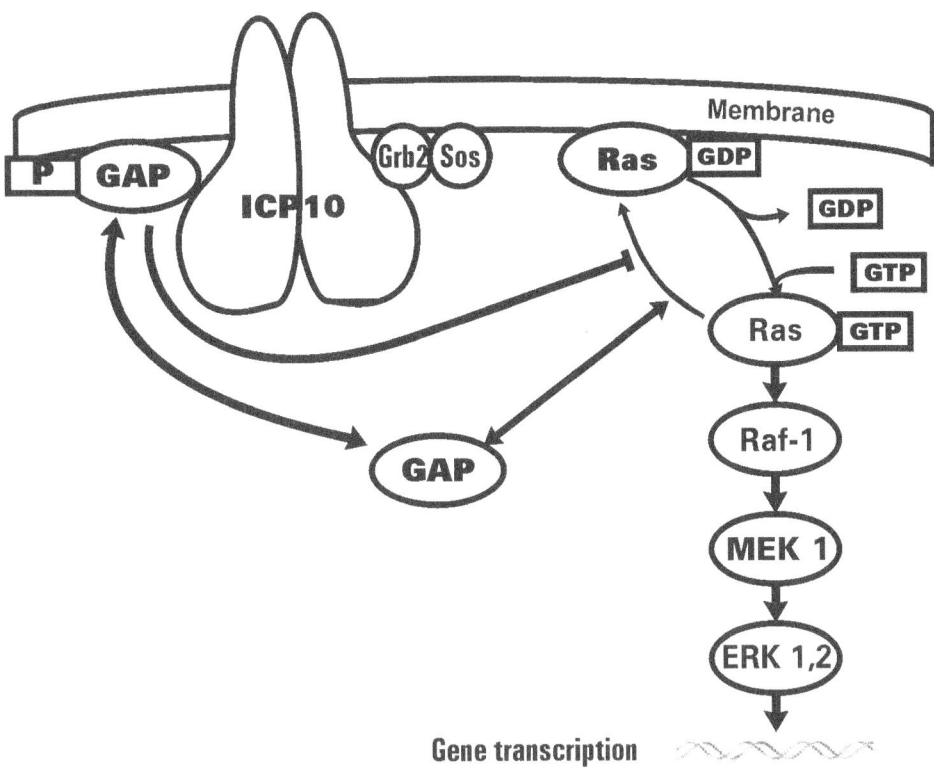

Fig. 14. **ICP10 activates the Ras/Raf/ERK pathway in non-neuronal cells.**

Consistent with its relative low homology to its HSV-2 counterpart, the HSV-1

RR1 PK (also known as ICP6 PK) is structurally and functionally distinct. It has different

ATP requirements and it lacks the conserved Lys of PK catalytic motif II (that

corresponds to Lys176 in ICP10 PK) (Peng *et al.*, 1996). The TM domain of ICP6 PK is

not followed by a negatively charged amino acid (required for TM anchorage) and

consequently ICP6 does not localize at the membrane (Aurelian, 1998). Unlike ICP10 PK which has both auto- and transphosphorylating activity (Chung *et al.*, 1989; Luo and Aurelian, 1992), ICP6 PK has only autophosphorylating activity (Conner *et al.*, 1992) and its biological role is not known.

Cellular proteins homologous to ICP10 PK such as FAST [a Fas-activated Ser/Thr kinase (Tian *et al.*, 1995)], and H11 [a HSP that is expressed in melanoma (Smith *et al.*, 1991)] have been identified, suggesting that ICP10 PK may have evolved from a cellular gene.

Modulation of Ras/Raf/MEK/ERK signaling pathway by ICP10 PK suggests that this viral protein may also be involved in modulation of apoptosis during HSV-2 infection. Significantly, the ICP10 PK cellular homologs were also implicated in apoptosis regulation (Tian *et al.*, 1995; Gober *et al.*, manuscript in preparation).

4.   Modulation of apoptosis by HSV

Both HSV-1 and HSV-2 have been shown to induce or protect cells from undergoing apoptosis. Induction of apoptosis occurs in a cell-type specific manner but the exact genes responsible have not yet been characterized. To date, no proteins with pro-apoptotic function were identified for HSV-1 or HSV-2, but both viruses were reported to induce apoptosis in certain cell types. The mechanisms by which wt HSV or various HSV mutants (i.e. deleted in ICP27, US3, ICP4/US3 or US5) induce apoptosis are not yet known. However, since they induce cell death in a cell type-dependent manner, it is possible that the putative pro-apoptotic viral genes require specific cellular factors and/or conditions.

Notwithstanding, a significant body of information has accumulated in the last years on the anti-apoptotic HSV genes, which are enumerated in Table 2. Although there is considerable evidence for the observed protective effect (which came mostly from studies of appropriate mutant viruses), the anti-apoptotic activity is cell type specific and its relationship to human disease is unclear.

Table 2

Virus	Genes	Anti-apoptotic mechanism
HSV-1	γ_1 34.5	binding/activation of PP1α
	ICP4	stabilization of Bcl-2
	US3	phosphorylation of Bad ?
	ICP27	stabilization of Bcl-2
	US5	inhibition of CD95/TNFR signaling ?
	LAT	?
HSV-2	US3	phosphorylation of Bad ?

As summarized in Table 2, the anti-apoptotic activity of HSV-1 was associated with a relatively large number of viral genes that have different functions and the expression of which is regulated with different kinetics. The ICP34.5 (γ_1 34.5) protein is important for viral replication (Bolovan *et al.*, 1994), viral ext from infected cells (Brown et al., 1994), neurovirulence (Chou *et al.*, 1990) and prevention of the premature shut-off of protein synthesis in the infected host (He *et al.*, 1997). Its anti-apoptotic mechanism is related to its binding/activation of PP1α, which dephosphorylates the translation initiation factor eIF2 (Whitley *et al.*, 1998), a process associated with protection from apoptosis induced by activated PKR. However, this interpretation is complicated by the fact that PP1α was also implicated in dephosphorylation/activation of the pro-apoptotic protein, Bad (as shown in Fig. 8). Significantly, a replication-restricted, HSV-1 mutant that lacks the ICP34.5 gene was able to kill p53-deficient ovarian cells that resist apoptosis and/or are chemotherapy resistant, suggesting a role for HSV-based oncolytic therapy in chemotherapy-resistant tumors

Two IE proteins (ICP27 and ICP4), which are known to regulate viral gene expression are also implicated in the anti-apoptotic activity of HSV-1. ICP27 is a multifunctional regulatory protein required for optimal expression of viral genes. Other functions of ICP27 include inhibition of RNA splicing and stimulation of DNA synthesis (Roizman and Sears, 1996). The second IE gene with anti-apoptotic activity is ICP4 (Roizman and Sears, 1996). ICP4 is believed to act both as an activator and repressor of viral genes. Its expression is required for the synthesis of most other viral proteins. Viral mutants that lack either ICP27 or ICP4 induce apoptosis in infected cells as evidenced by activation of caspases, cyt c release and DNA fragmentation. Apoptosis was associated with downregulation of Bcl-2 levels by destabilization of its RNA, decreased half-life of the protein and caspase-mediated degradation of Bcl-2. Wild-type HSV-1 did not cause apoptosis in these cells and stabilized the Bcl-2 RNA and protein levels, suggesting that ICP27 and ICP4 have anti-apoptotic activities that involve stabilization of Bcl-2 (Aubert and Blaho, 1999; Aubert *et al.*, 1999; Zachos *et al.*, 2001). Significantly, even though the p38MAPK and JNK pathways were activated by infection with wt or the two mutants, pathways inhibition correlated with apoptosis (specifically with inhibition of cell death) only in mutant-infected cells (Zachos *et al.*, 2001). By contrast, inhibition of p38MAPK in wt HSV-1-infected cells resulted only in decreased viral yield, supporting previous data that suggested a role for stress-activated pathways in enhancing virus replication, and in apoptosis in some but not all cell types (Zachos *et al.*, 1999). Leopardi and Roizman (1996) also reported that an ICP4 deleted HSV-1 mutant induces apoptosis, but the mutant was later shown to also have a defect in the US3 gene which codes for a Ser/Thr PK (Galvan and Roizman, 1998; Galvan *et al.*, 1999). Further studies with this double mutant (ICP4 and US3-deleted) showed that it induces caspase-independent apoptosis (Galvan and Roizman, 1998; Galvan *et al.*, 1999). However, in HSV-1 infected cells exposed to osmotic shock (sorbitol) caspase-3 was activated, whereas the MPT as well as the cyt c release and DNA fragmentation were blocked, suggesting that HSV-1 has anti-apoptotic functions that act on both caspase-dependent and independent pathways. By contrast, in other cells infected with the double mutant, apoptosis was efficiently blocked by caspase inhibitors and overexpression of Bcl-2 suggesting that distinct apoptotic pathways are induced (and blocked) in different cell types. HSV-1 US3 was also involved in protection from apoptosis induced by UV, Fas, thermal or osmotic shock (Leopardi *et al.*, 1997; Jerome *et al.*, 1999). Its mechanism of action may involve phosphorylation/inactivation of the pro-apoptotic protein Bad, since US3 was shown to phosphorylate Bad, in an *in vitro* system involving co-expression of the two proteins (Munger and Roizman, 2001); yet, it remains to be seen if indeed this occurs in the case of HSV-1 infection. Interestingly, even though US3 is able to inhibit apoptosis induced by ICP4/US3 double mutant (as evidenced by inhibition of cyt c release and caspase-3 activation), an HSV-1 mutant deleted in the US3 does not induce apoptosis, suggesting that other genes may compensate and induce survival (Munger *et al.*, 2001).

The US5 gene codes for the viral glycoprotein J (McGeoch, 1990). US5 protected cells against Fas-induced apoptosis and partially protected against UV-induced apoptosis by an yet uncharacterized pathway which may involve interference with signaling initiated from the death receptor (e.g. CD95/Fas) (Jerome *et al.*, 1999).

The latency associated transcript (LAT), the only viral gene transcribed in latently infected ganglia is a stable intron (2 kb) derived by splicing from a primary RNA transcript of 8 kb (Roizman and Sears, 1996). Extensive apoptosis was seen in trigeminal ganglia of rabbits infected with a LAT-deleted mutant, suggesting that LAT has anti-apoptotic activity (Perng *et al.*, 2000). LAT was also implicated in protection from apoptosis induced by C6-ceramide and etoposide in transiently transfected cells (Perng *et al.*, 2000). Because these two stimuli act by two different pathways, LAT is likely to act on a common apoptosis effector. The anti-apoptotic activity of LAT may be biologically important during reactivation of latent virus since reactivating stimuli (e.g. corticosteroid hormones) are also involved in triggering apoptosis (Dieken and Miesfield, 1992).

However, the role of LAT in the anti-apoptotic activity of HSV is difficult to explain in the context of the present knowledge of signaling and apoptotic pathways which mainly involve proteins.

In contrast to the anti-apoptotic genes encoded by the HSV-1, the anti-apoptotic genes encoded by HSV-2 (and described up to date) are not well characterized. The anti-apoptotic HSV-2 US3 protein has a 75 % level of identity to its HSV-1 counterpart and, like its HSV-1 counterpart, was shown to have anti-apoptotic activity. An US3-deleted mutant induced DNA fragmentation early in infection, in contrast to wt HSV-2, suggesting that US3 is responsible for the wt virus anti-apoptotic activity. However, the mutant did not induce apoptosis at a later time suggesting that late in infection another viral anti-apoptotic function may also be activated (Hata *et al.*, 1999). *In vivo* studies (Asano *et al.*, 1999) of corneal infection in mice showed that US3 deleted mutant induces apoptosis whereas the wild type and revertant virus do not, suggesting an *in vivo* anti-apoptotic role of US3 protein kinase. The mechanism of HSV-2 US3 anti-apoptotic activity is not yet known but it seems plausible (due to the high homology to its HSV-1 counterpart) that it also acts by phosphorylating/inhibiting the pro-apoptotic protein Bad.

In conclusion, HSV-1 has multiple anti-apoptotic genes that act in a cell-type and stimulus specific manner. Their HSV-2 counterparts may also have anti-apoptotic activity but it was specifically demonstrated only for US3. The only HSV gene shown to inhibit cell death in neurons is LAT (Perng *et al.*, 2000), but the exact mechanism involved is not yet clear and it remains a challenge to understand how an RNA transcript is able to modulate neuronal apoptosis.

D.   Specific aims

The different outcome of CNS infection with HSV-1 and HSV-2 suggests that HSV-2 has specific anti-apoptotic gene(s) that act in the CNS, specifically in the hippocampus, which is the area most damaged in the HSE. Among the putative anti-apoptotic genes, ICP10 PK is a potential candidate in virtue of its ability to activate the Ras/Raf/MEK/ERK pathway in non-neuronal cells (Smith *et al.*, 2000) which was associated with modulation of apoptosis (Kaplan and Miller, 2000). Identification of a viral gene with anti-apoptotic activity in hippocampal neurons may lead to the development of novel therapeutic strategies for treatment of neurological disorders with an apoptotic determinant. The present studies were designed to test this hypothesis.

The specific aims are :

A.   To examine whether ICP10 PK has anti-apoptotic activity in non-neuronal and neuronal cells.

B.   To investigate the targets of HSV-2-induced survival activity in hippocampal neurons (activation of Raf/MEK/ERK and/or PI3-K/Akt pathways, inhibition of pro-apoptotic JNK signaling pathway, and the respective transcription factors involved (CREB, fos, c-Jun, ATF-2, p53) and to determine the effect of ICP10 PK on Bcl-2 family of proteins (Bcl-2, Bad) and their chaperones (e.g. Bag-1).

C.   To examine the cell death mechanism in ICP10ΔPK- and HSV-1-infected hippocampal neurons, and HSE-affected human brain.

D. To determine whether ICP10 PK protects from apoptosis induced by stimuli specific for neurological disorders (i.e. growth factor withdrawal, oxidative stress or genetic defects that trigger apoptosis).

E. To establish a murine model of ICP10 PK gene delivery to the brain, using intranasal infection with a growth-deficient HSV-2 mutant that expresses ICP10 PK.

II   MATERIALS AND METHODS

Cells. Vero (African green monkey kidney) and HEp-2 cells were grown in Minimum Essential Medium (MEM) with 10 % fetal bovine serum (FBS) and 100 U/ml Penicillin-Streptomycin (Gibco-BRL, Gaithersburg, MD). NIH3T3 cells were the gift of Dr. Paul Anderson (Brigham and Women's Hospital, Harvard Medical School, Boston, MA) and were grown in DMEM, 10 % FBS and 100 U/ml Penicillin-Streptomycin. Mouse neuroblastoma N2a cells stably transfected with wild-type (WT) and mutant (G85R) SOD1 cDNA were obtained from Dr. David R. Borchelt (Department of Pathology, The Johns Hopkins School of Medicine, Baltimore, MD). They were grown in 50 % DMEM and 50 % OPTI-MEM with 5 % FBS and 400 μg/ml G418 and differentiated for 6 days in serum free medium as described (Pasinelli *et al.*, 1998; Pasinelli *et al.*, 2000). Rat pheochromocytoma (PC12) cells were obtained from Dr. Paul Yarowsky (Department of Pharmacology and Experimental Therapeutics, University of Maryland, Baltimore, MD) and cultured in DMEM/F12 (Gibco-BRL) with 10 % FBS, 0.36 % D-glucose (Sigma), 0.21 % sodium bicarbonate (Sigma) and 0.009 % gentamycin (Sigma). They were neuronally differentiated by growth (at least 12 days) in serum-free DMEM/F12 supplemented with 0.36 % D-glucose, 0.21 % sodium bicarbonate, 0.009 % gentamycin and 100 ng/ml nerve growth factor (NGF) (Roche Molecular Biochemicals). Cell lines JHLa1 and JHL15 were established from human embryonic kidney cells

(HEK293). They respectively express ICP10 PK or its kinase negative mutant p139^{TM} (Smith et al., 1994). JHLa1, JHL15 and HEK293 cells were grown in Dulbecco's Modified Eagle's Medium (DMEM) with 10 % FBS, 1 mM sodium pyruvate, 0.1 mM non-essential amino acids and 100U/ml Penicillin-Streptomycin (Gibco-BRL).

Establishment of primary neuronal cultures. Primary cultures of cells dissociated from the hippocampi or cerebral cortex of 16- to 19-day old rat fetuses (Sprague-Dawley) were prepared according to the procedure described by Alkondon and Albuquerque (1993). Cells were plated at a density of approximately 750,000/2ml on collagen coated 35mm dishes (Nunc, Rochester, NY) or glass coverslips precoated with poly-L-Lysine (Sigma, St. Louis, MO). The studies described in this report were done with 6-day old cultures which consisted mostly (>85 %) of non-dividing cells [as determined with the 5-Bromo-2'-deoxyuridine labeling and detection kit (Roche Molecular Biochemicals, Indianapolis, IN)], identified as neurons by immunocytochemistry with neuron-specific antibodies [neurofilament-160 kDa (NF-160) and class III beta-tubulin (TUJ1)]. To generate trisomy 16 (Ts16) mice, male mice doubly heterozygous for Robertsonian translocations of chromosome 16 [Rb(6,16)24LuBXRb(16,17)7BNRF1] were mated with C57BL/6J female mice. Primary hippocampal cultures were established from embryonic (day 16) euploid and Ts16 fetuses [distinguished from one another and karyotyped as previously described (Haydar *et al.*, 2000)]. Cells were grown on glass coverslips etched with a grid of 175 x 175 μm squares (CELLocate; Eppendorf, Madison, WI) in MEM with B27 supplement (Gibco) which contains optimized concentrations of neuron survival factors, as previously described

(Bambrick *et al.*, 1995; Bambrick and Krueger, 1999). Neuronal identity (> 90 %) was confirmed by staining with the neuron-specific TuJ1 antibody (Ferreira and Caceres, 1992). 2-day old primary mouse hippocampal cultures were used for these studies.

Brain samples. Post-mortem human brain samples (temporal lobe) of 8 HSE patients, and 2 controls were obtained from Dr. Kym Gyure, Department of Pathology, University of Maryland, Baltimore, MD). Brain tissue was fixed in 10 % formalin (pH 7.6) and embedded in paraffin. The HSE diagnosis had been confirmed by the presence of characteristic Cowdry A intranuclear inclusions (in neurons, glia and endothelial cell nuclei) in the setting of encephalitis, positive immunohistochemical stains for HSV-1, and/or cultures positive for HSV-1. Sections were negative for HSV-2-specific antigens by these methods. All HSE cases were evaluated with hematoxylin-eosin staining and showed varying degrees of viral encephalitis characterized by perivascular lymphoid infiltrates, microglial nodules and neuronophagia. In addition, most cases additionally demonstrated foci of hemorrhage/subacute hemorrhagic necrosis with macrophage-rich inflammatory infiltrate. The control (normal) human brains were from individuals with no neurological or psychiatric history.

Brains from C57BL/6J mice infected intranasally with ICP10ΔRR (5 μl of virus suspension containing 5 x 10^6 pfu/nostril) or mock infected with PBS (5 μl/nostril), were infused with 20 % sucrose, frozen and cryostat sections were cut in a coronal plane from the olfactory bulbs through the hippocampus.

Viruses and cell infection. Wild type viruses [HSV-2 (strain G) and HSV-1 (strain F)] and the HSV-2 mutants [ICP10ΔPK and ICP10ΔRR (that are respectively deleted in the PK or RR domains of ICP10)] were previously described (Peng *et al.*, 1996; Smith *et al.*, 1998; Aurelian and Smith, 2000). ICP10ΔPK and ICP10ΔRR were differentiated from HSV-2 by plaque morphology and immunoblotting with ICP10 antibody (recognizes the PK-negative 95 kDa protein in ICP10ΔPK, the p175 kDa ICP10 PK- *LacZ* fusion protein in ICP10ΔRR, and the 140 kDa ICP10 protein in HSV-2). The ICP10 PK gene was restored in the ICP10ΔPK mutant to yield the revertant virus [HSV-2 (R)] that expresses the 140 kDa ICP10 protein, as previously described (Smith *et al.*, 1998). HSV-1 mutants ICP6Δ (lacks 90 % of the coding sequence of the ICP6 gene) and *hr*R3 (retains only 38 % of the N-terminus domain of ICP6 and has the *LacZ* gene inserted in the RR domain) were kindly provided by Dr. Sandra K. Weller (University of Connecticut Health Center, Farmington, CT) and were previously described (Goldstein and Weller, 1987; Goldstein and Weller, 1988).

All viruses were grown on Vero cells with MEM and 10 % FBS. Virus titers were determined by plaque assay under an overlay consisting of MEM supplemented with 0.3 % human serum IgG and 10 % FBS, as described (Aurelian, 2000). All viruses used throughout these studies had similar titers [5 x 10^8 - 1 x 10^9 plaque forming units (pfu)/ml]. The experimental design accounted for similar multiplicity of infection (m.o.i.; which is defined as number of pfu/cell) and same volume of viral suspension and growth medium-as control.

Infection of neuronal cultures was done at a multiplicity of infection (m.o.i.) of 10 for all experiments except the one-step single growth curves. After the adsorption phase (1 hr, 36.5^0 C) virus inoculum was removed and cultures were overlaid with growth medium containing 10 % serum- a condition that allows ICP10ΔPK and ICP10ΔRR replication in non-neuronal cells (Smith *et al.*, 1994; Aurelian and Smith, 2000).

For UV inactivation, virus [100μl; 1 x 10^7 plaque forming units (pfu)] was placed in a 24-well plate and exposed to UV irradiation for 45 min using a Sylvania G15 T8 bulb at a distance of 17 cm with occasional agitation (Purifoy and Powell, 1977). For neutralization with the anti-HSV-2 or preimmune sera, HSV-2 (10^7 pfu) was incubated (1 hr, 37^0C) with an equal volume of the respective sera.

For neutralization with the gD MAb, cells were pre-incubated for 30 min with 2.5 g/ml of the IgG fraction and exposed (1 hr; 37^0C) to a virus-antibody mixture consisting of HSV-2 (10^7 pfu) and 2.5 g/ml of gD IgG (1 hr; 37^0C). Virus surviving neutralization was determined by plaque assay as described (Aurelian, 2000).

Mouse intranasal infection. C57BL/6J mice (6-8 weeks of age) were anaesthetized with ether and infected by expelling 5 μl of ICP10ΔRR virus suspension (5 x 10^6 pfu) from a micropipettor into each nostril. PBS (5 μl/nostril) was used as control.

Antibodies. Polyclonal antibody to ICP10 (recognizes amino acids 13-26) was previously described (Aurelian *et al.*, 1989). It recognizes ICP10 and the p95 and p175 proteins respectively expressed by ICP10ΔPK and ICP10ΔRR (Aurelian and Smith, 2000) The hyperimmune HSV-2 serum was generated in mice injected with HSV-2 (10^6 pfu) as described (Aurelian *et al.*, 1999). Antibody to class III beta-tubulin (TUJ1) was the gift of Dr. Paul Yarowsky (University of Maryland, Baltimore, MD). The following antibodies were purchased and used according to the manufacturer's instructions: monoclonal antibodies (MAbs) to neurofilament 160 (NF-160), glial fibrillary acidic protein (GFAP) and galactocerebroside (GalC) [Neural Cell Typing Set for Identification and Typing of Neural Cells (Roche Molecular Biochemicals)], the neutralizing IgG fraction of a MAb to HSV-1/2 glycoprotein D (gD) (Advanced Biotechnologies, Columbia, MD) and the monoclonal antibody to β-gal (Sigma); polyclonal antibodies to cleaved caspase-3 [D175, recognizes the activated (caspase3p20) species] (Cell Signaling Technology, Beverly, MA), FLAG (Cayman Chemicals, Ann Arbor, MI), PARP p85 fragment [recognizes the 85 kDa caspase-cleaved fragment of PARP (Promega, Madison, WI)], active JNK [recognizes the dually phosphorylated, active form of JNK (Promega)], JNK [reacts with JNK1, JNK2 and JNK3 (Santa Cruz Biotechnology, Santa Cruz, CA)], actin (Santa Cruz Biotechnology), c-fos (Santa Cruz Biotechnology), Bcl-2 (ΔC21, Santa Cruz Biotechnology), Bag-1 (FL-274, Santa Cruz Biotechnology), ERK (recognizes ERK1/2) (Oncogene, Cambridge, MA), active ERK (recognizes the dually phosphorylated active forms of ERK1/2) (P-ERK1/2) (Promega Corporation, Madison, WI), PARP [recognizes the uncleaved (112 kDa) and cleaved (85 kDa) PARP] (Roche Molecular Biochemicals), caspase-3 [recognizes the uncleaved (caspase3p32) and cleaved (caspase3p20) species] (Santa Cruz Biotechnology), Raf-B (reacts with p68 and p95 species of Raf-B) (Santa Cruz Biotechnology), Bad (Calbiochem), p53 (Ab-7, Oncogene). Also, polyclonal antibodies to phospho-c-Jun (Ser63), phospho-c-Jun (Ser 73) and c-Jun, were part of the PhosphoPlus c-Jun (Ser63)II c-Jun (Ser73) Antibody Kit (Cell Signaling Technology); polyclonal antibodies to phospho-CREB (Ser 133) and CREB-part of the PhosphoPlus CREB (Ser133) Antibody Kit (Cell Signaling Technology), and polyclonal antibodies to phospho-ATF-2 (Thr69/71) and ATF-2 (Cell

Signaling Technology) were the gift of Dr. Biswendu Goswami (Food and Drug Administration, Washington, D.C.).

Pharmacological inhibitors. The PI3-K inhibitors wortmannin (Yano *et al.*, 1998) and LY294002 (Vlahos *et al.*, 1994) were respectively purchased from Sigma and Calbiochem (San Diego, CA). c-Raf-1 kinase inhibitor I (Lackey *et al.*, 2000) and p38 MAPK inhibitor (Cuenda *et al.*, 1995) were purchased from Calbiochem (San Diego, CA). The MEK-specific inhibitor U0126 (Favata *et al.*, 1998) was from Promega. The AMPA/KA receptor inhibitor amino-5-phosphonovaleric acid (APV) (Sigma), NMDA receptor inhibitor 6-cyano-7-nitroquinoxaline-2,3-dione (CNQX) (Research Biomedicals, Natick, MA) and acetylcholine receptor antagonist α-bungarotoxin (Biotoxins, Inc., St. Cloud, FL) were the gift of Dr. Edna Pereira (Department of Pharmacology and Experimental Therapeutics, University of Maryland, Baltimore). Other reagents used were from Sigma (D-mannitol, sorbitol and xanthine), Roche Molecular Biochemicals (xanthine oxidase) and Calbiochem (staurosporine).

Plasmids and transfection. The Bag-1 expression vector (pJG4-5 m Bag-1) that encodes the mouse Bag-1 protein was purchased from Science Reagents Inc. (Atlanta, GA) and was previously described (Takayama *et al.*, 1997). The c-Raf-1 K375M-FLAG dominant negative plasmid was kindly provided by Dr. Bernard Weinstein, Columbia University, NY. Its construction and characterization have been previously reported (Dent *et al.*, 1995; Soh *et al.*, 1999). The construction of pJW17 and pJHL15 expression vectors that respectively express ICP10 (pJW17) or a TM deleted PK negative ICP10 mutant called p139[TM] (pJHL15) was previously described (Chung *et al.*, 1989; Luo and Aurelian,

1992). Transfection was done with FuGene 6 Transfection Reagent (Roche Molecular Biochemicals) according to the manufacturer instructions. The efficiency of transfection in neuronally differentiated PC12 cells and primary hippocampal cultures was between 30-40 % for all plasmids, as described in the text.

TUNEL (TdT-mediated dUTP nick end labeling). Cells treated with various apoptotic inducers or infected with HSV-1, HSV-2, ICP10ΔPK or ICP10ΔRR or mock-infected with growth medium, were examined for apoptosis using the *In Situ* Cell Death Detection Kit (Roche Molecular Biochemicals) according to the manufacturer's instructions. Briefly, cells were fixed with 4 % paraformaldehyde (PFA) in PBS (pH 7.4) for 1 hr at room temperature followed by permeabilization in 0.1 % Triton X-100 (in 0.1 % sodium citrate) for 2 minutes on ice. DNA breaks were labeled by addition of terminal deoxynucleotidyl transferase (TdT) and nucleotide mixture [containing fluorescein (FITC)- conjugated dUTP] and incubation for 60 minutes at 37° C. Coverslips were mounted in PBS/glycerol and cells were analyzed by fluorescence microscopy. After extensive washes in PBS, cells were incubated for 30 minutes at 37°C with an anti-FITC antibody conjugated with alkaline phosphatase (AP). Chromogenic reaction was carried out by adding AP substrate solution [0.4 mg/ml nitro blue tetrazolium chloride (NBT) and 0.2 mg/ml 5-bromo-4-chloro-3-indolyl phosphate, toluidine salt (BCIP) (Roche Molecular Biochemicals) in 0.1 M Tris-HCl pH 9.5, 0.05 M $MgCl_2$, 0.1 M NaCl and 1 mM levamisole] for 10 minutes at room temperature. Coverslips were mounted in PBS/glycerol and analyzed by light microscopy. Apoptotic cells (characterized by a dark nuclear precipitate) and non-apoptotic cells (unstained or displaying a diffuse, light and

96

uneven cytoplasmic staining) were counted in 5 randomly chosen microscopic fields (containing at least 250 cells). Results are expressed as % apoptotic cells $\pm$ SEM.

Viability (MTS) assay. Cell viability was determined using CellTiter 96 Aqueous One Solution Cell Proliferation Assay (Promega) according to the manufacturer instructions. Briefly, the levels of cellular [3-{4,5-dimethylthiazol-2-yl)-5-(3-carboxymethoxyphenyl)-2-(4-sulphophenyl)-2H-tetrazolium, inner salt] (MTS) reduction to formazan, a measure of mitochondrial function, were quantified by adding the CellTiter 96 Aqueous One Solution (1:5; v:v; MTS solution/culture medium) to the cells grown in 96 well plates. After 1 hr at 37^0 C, the absorbance at 490 nm was read using an ELISA plate reader. Results were confirmed to fall within the linear portion of the titrated cell number: absorbance curves. Values are the mean $\pm$ SEM of three independent experiments (each done in quadruplicates) and are expressed as % survival (MTS reduction) of cells at 0 hr post-treatment $\pm$ SEM.

Caspase activity assay. Caspase activation was measured by CaspACE FITC-VAD-FMK *In Situ* Marker (Promega) according to the manufacturer's instructions. Briefly, hippocampal cultures infected with HSV-1, ICP10ΔPK, ICP10ΔRR or HSV-2 at m.o.i. 10 or mock-infected for 24 hrs were treated with the FITC conjugate of the permeable irreversible caspase inhibitor (VAD-FMK) for 20 min to allow binding to activated caspases. Cells were subsequently fixed with 10 % buffered formaldehyde (1 hr, room temperature) and visualized by fluorescence microscopy.

Immunofluorescent staining. The identity of the TUNEL-positive cells in primary hippocampal cultures was determined by double immunofluorescence. The cultures were incubated with TdT and the nucleotide mixture (containing FITC-conjugated dUTP) for 1 hour at 37^0C and stained (1 hour, room temperature) with antibodies NF-160, TUJ1, GFAP or GalC followed by phycoerythrin (PE)-conjugated anti-mouse IgG (30 min, room temperature). Stained cells were visualized with an epifluorescent confocal microscope fitted with an argon ion laser (Zeiss LSM 410) as described (Aurelian *et al.*, 2001).

Immunocytochemistry. The DAKO LSAB 2 Kit, HRP (DAKO Corporation, Carpinteria, CA) was used for immunoperoxidase staining. Cells were exposed overnight (4^0 C) to primary antibodies (caspase-3p20, active JNK, FLAG or p85 PARP) and immunolabeled cells were subsequently detected using the streptavidin-biotin method according to the manufacturer's instructions. Counterstaining was with Mayer's hematoxylin (Sigma) (Kokuba *et al.*, 1999). Five randomly chosen microscopic fields (containing at least 250 cells) were counted and the percentage of mock-infected positive cells was subtracted from each average. Results are expressed as % positive cells $\pm$ SEM..

Immunohistochemistry. The DAKO LSAB 2 Kit, HRP (DAKO Corporation, Carpinteria, CA) was used for immunoperoxidase staining of paraffin-embedded or cryofrozen brain tissue. Paraffin-embedded brain tissue was deparaffinized through a series of xylene-ethanol washes. Sections were exposed overnight (4^0 C) to primary antibodies (caspase-3p20, active JNK, p85PARP or ICP10) and immunolabeled cells were subsequently detected using the streptavidin-biotin method according to the manufacturer's instructions. Counterstaining was with Mayer's hematoxylin (Sigma). At least 300 cells were counted and results were expressed as % positive cells ± SEM.

X-gal staining. Vero cells (grown with MEM, 10 % FBS) were infected with ICP10ΔRR virus at m.o.i. 10. At 24 hrs p.i. cells were stained by adding 400 μg of the chromogenic substrate X-gal (5-bromo-4-chloro-3-indolyl-β-D-galactoside) per ml of medium and visualized within 12 to 24 hrs.

Hoechst staining. Cells were fixed in 4 % PFA (pH 7.4) for 20 min and stained with 1 μg/ml of Hoechst 32258 (Sigma) in PBS for 10 minutes, as previously described (Guo *et al.*, 1997).

DNA fragmentation. DNA fragmentation was assayed as described by Hata *et al.*, (1999). Cells were collected by trypsinization, DNA was extracted as described by Hirt (1967), quantitated by spectrophotometry and 5-10 μg were separated on 1.5 % agarose gels. Gels were stained with 0.1 g/ml ethidium bromide and visualized by exposure to UV light. Photographs were taken with a Polaroid camera (Kodak).

Immunoblotting assay. Immunoblotting was done as previously described (Smith *et al.*, 2000). Briefly, cells were lysed with RIPA buffer (30 mM Tris-HCl pH 7.4, 0.15 mM NaCl, 1 % Nonidet P-40, 0.1 % SDS, 0.5 % sodium deoxycholate, 1 mM EDTA, 1 mM DTT, 2 mM MgCl$_2$, 0.5 mM PMSF) supplemented with phosphatase and protease inhibitors cocktails (Sigma) and sonicated for 30 seconds at 25 % output power using the Sonicator/Ultrasonic Processor (Misonix, Inc., Farmingdale, NY). Total protein was determined by the bicinchoninic assay (Pierce, Rockford, IL) and proteins were resolved by SDS-PAGE and transferred to nitrocellulose membranes. The blots were incubated (1 hr, 37^{0}C) in TN-T buffer (0.01 M Tris- HCl pH 7.4, 0.15 M NaCl, 0.05 %

Tween 20) containing 1 % bovine serum albumin (BSA) to block non-specific binding and exposed (2 hrs, room temperature) to the appropriate antibodies (diluted in TN-T buffer with 0.1 % BSA). After 3 washes with TN-T buffer the blots were incubated with Protein A-Peroxidase for 1 hr at room temperature. Detection was with ECL reagents (Amersham Life Science, Arlington Heights, IL) and exposure to high performance chemiluminescence film (Hyperfilm ECL, Amersham). Quantitation was by densitometric scanning using the BioRad GS-700 Imaging Densitometer. All immunoblotting experiments were repeated at least 3 times for each antibody. A representative blot is shown for each result.

Single-step growth assays. Primary cultures of hippocampal neurons (day 6 in culture) were infected with HSV-2, HSV-1, ICP10ΔPK, or ICP10ΔRR at m.o.i. 5. Adsorption was done for 1 hr (0 hr in the growth curve). Cells and supernatants were harvested at various times between 0 and 48 hrs p.i., frozen, thawed and assayed for virus titers by plaque assay (Aurelian, 2000). Results are expressed as mean pfu/ml $\pm$ SEM.

Statistical analyses. Student t test and ANOVA with Tukey-Kramer post-test were performed using GraphPad InStat version 3.01 for Windows 95/NT, GraphPad Software, San Diego, CA.

III   **RESULTS**

    Specific Aim 1. To examine whether ICP10 PK has anti-apoptotic activity in non-neuronal and neuronal cells.

    <u>In a preliminary series of experiments, the putative anti-apoptotic activity of ICP10 PK was tested in human cells derived from HEK293 that were stably transfected with ICP10, and treated with various apoptotic inducers.</u>

1.   **ICP10 PK protects from apoptosis induced by staurosporine (STS) or D-mannitol (D-Mann).** It was previously shown that the Ras/Raf/MEK/ERK mitogenic pathway is activated in HEK293 cells that constitutively express ICP10 PK (JHLa1), but not in cells that express its PK negative mutant, p139[TM] (JHL15) (Smith *et al.*, 1994). Because the Ras/Raf/MEK/ERK pathway was implicated in the control of apoptosis (Kaplan and Miller, 2000), the response of JHLa1 and JHL15 to apoptotic inducers was examined and compared to untransfected, parental HEK293 cells. Staurosporine (STS) and D-mannitol (D-Mann) were chosen as apoptotic inducers because they act by different pathways, by inhibiting cellular kinases (STS) or by inducing osmotic shock (D-Mann) (Jacobson *et al.*, 1996; Malek *et al.*, 1998). Cells were treated for 24 hrs with STS (250 nM) or D-Mann (300 mM) in culture medium containing 1 % FBS and examined by TdT-mediated dUTP nick end labeling (TUNEL), an assay that is widely considered to be

specific for apoptosis (Gavrieli *et al.*, 1992; Gold *et al.*, 1994). Apoptotic cells (characterized by a dark nuclear precipitate) and non-apoptotic cells (unstained or displaying a diffuse, light and uneven cytoplasmic staining) were counted in 5 randomly chosen microscopic fields (containing at least 250 cells). Results of three independent experiments are expressed as % apoptotic cells $\pm$ SEM. The proportion of TUNEL-positive cells (apoptotic) was significantly higher in STS treated HEK293 (71 ± 12.4 %) and JHL15 (69 ± 4.9 %), than JHLa1 (5.8 ± 2.1 %) cells, and similar results were obtained for D-Mann (79 ± 7.4 %, 58 ± 7 % and 13 ± 8 % for HEK293, JHL15 and JHLa1 cells respectively) ($p<0.01$ by Student t test) (Fig. 15 A). TUNEL-positive (STS-treated) HEK293 and JHL15 cells evidenced the hallmark morphological features of apoptosis including cell shrinkage, condensed chromatin and nuclear fragmentation (Fig. 15 B, C). These features were not seen in the TUNEL negative (STS-treated) JHLa1 cells (Fig. 15D). The data indicate that ICP10 PK has anti-apoptotic potential because ICP10 PK activity is the only difference between JHLa1 and JHL15 cells (Smith *et al.*, 1994).

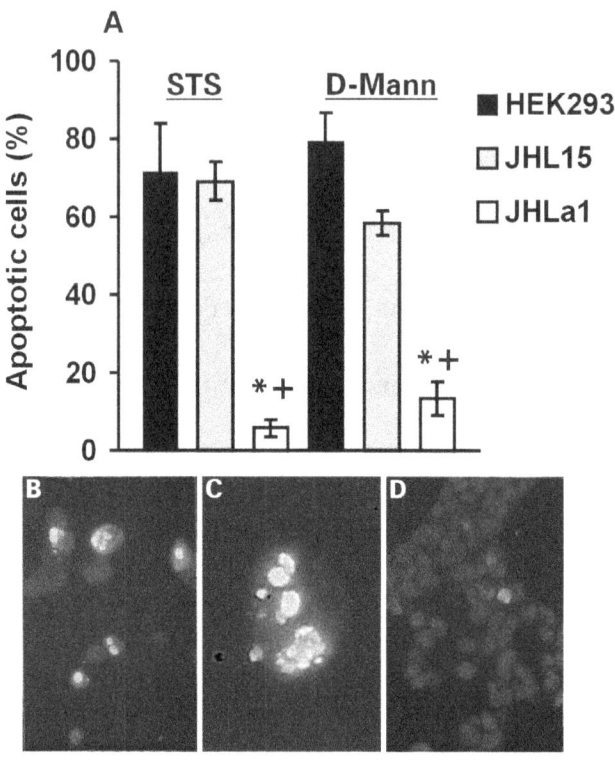

Fig. 15.   ICP10 PK protects from STS and D-Mann-induced apoptosis. (A)

HEK293, JHL15 and JHLa1 cells were treated with 250 nM STS or 300 mM D-Mann for

24 hrs and analyzed for cell death by TUNEL as described in Materials and Methods.

Apoptotic (TUNEL-positive) and non-apoptotic (TUNEL-negative) cells were counted in

5 randomly chosen microscopic fields. The results of three independent experiments are

expressed as mean % apoptotic cells $\pm$ SEM. (* = p < 0.01 vs. HEK293 ; + = p < 0.01

vs. JHL15 by Student t test).**(B,C,D)** HEK293 (B) and JHL15 cells (C) treated with STS-

treated and labeled by TUNEL evidenced typical apoptotic nuclear morphology, which

was not seen in JHLa1 cells (D)

2. **ICP10 PK blocks STS or D-Mann induced DNA fragmentation.** The ability

of ICP10 PK to block DNA fragmentation was tested next, because DNA fragmentation

is a hallmark of most, albeit not all, apoptotic conditions (Cohen *et al.*, 1992). DNA

fragmentation was assayed as described by Hata *et al.* (1999). Briefly, DNA was

extracted from HEK293 and JHLa1 cells that were treated for 24 hrs with STS (250 nM)

or D-Mann (300 mM) as described by Hirt (1967) and quantitated by spectrophotometry.

DNA (5-10 µg) was separated on agarose gel (1.5 %), stained with 0.1 µg/ml ethidium

bromide and visualized by exposure to UV light. Cells mock-treated with DMSO (solvent

for STS) or MEM (solvent for D-Mann) were studied in parallel. The degradation of

chromosomal DNA into oligonucleosomal fragments, a characteristic of apoptosis was

seen in HEK 293 cells treated with STS (Fig. 16, lane 3) or D-Mann (Fig. 16, lane 7) but

not in mock-treated HEK293 cells (Fig. 16, lanes 2 and 6). Similar results were obtained

for JHL15 cells (data not shown). However, DNA fragmentation was not seen in JHLa1

cells treated with STS (Fig. 16, lane 4) or D-Mann (Fig. 16, lane 8), indicating that ICP10

PK blocks DNA fragmentation induced by apoptotic stimuli.

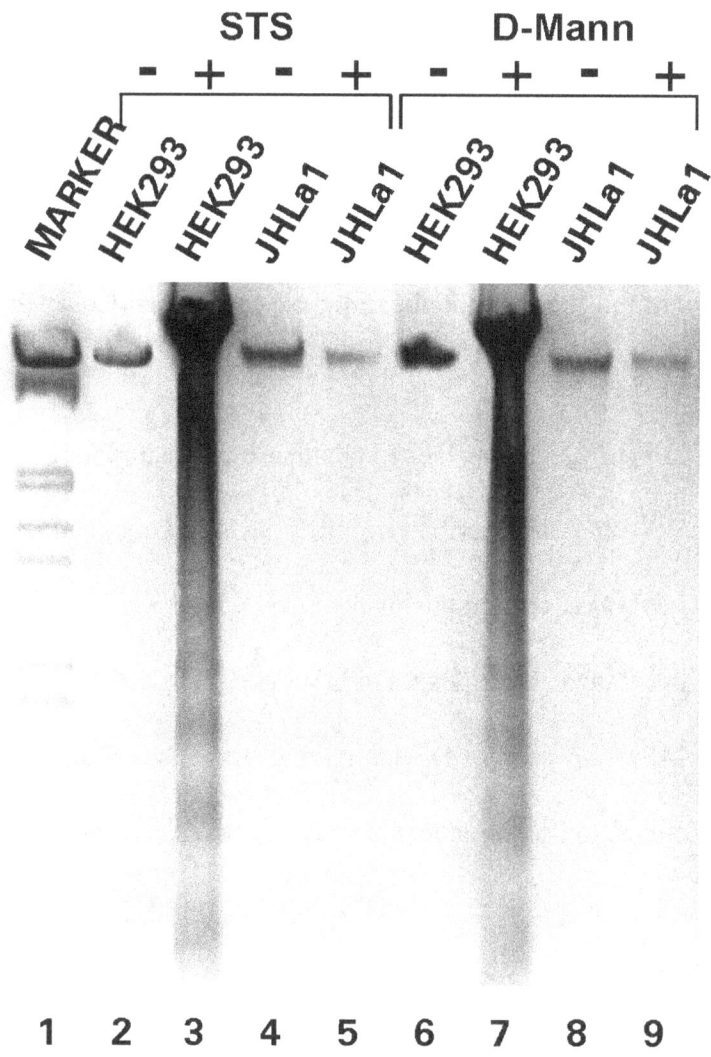

Fig. 16. **ICP10 PK inhibits DNA fragmentation.** HEK293 (lanes 2, 3, 6, and 7) and

JHLa1 (lanes 4, 5, 8, and 9) cells were treated with 250 nM STS (lanes 3 and 4) or 300

mM D-Mann (lanes 7 and 8) or mock treated with DMSO (control for STS) (lanes 2 and

5) or MEM (control for D-Mann) (lanes 6 and 9). Genomic DNA was extracted at 24 hrs

of treatment and separated on agarose gels as described in Materials and Methods. Lanes

are identified by numbers at the bottom of the figure.

3. **Apoptosis inhibited by ICP10 PK in JHLa1 cells is caspase-3-dependent.** To

further examine the anti-apoptotic activity of ICP10 PK, its ability to inhibit cleavage of

pro-caspase-3, a central determinant of many apoptotic processes (Johnson-Webb *et al.*,

1997), was tested in two series of experiments. In the first series, it was determined

whether ICP10 PK interferes with cleavage of the proenzyme (caspase-3p32). A 32 kDa

band consistent with caspase-3p32 was seen by immunoblotting with caspase-3p32

antibody in HEK293, JHL15 and JHla1 cells treated with DMSO (STS diluent) or STS

250 nM. However, densitometric scanning indicated that its levels were significantly

lower in HEK293 and JHL15 cells treated with STS (Fig. 17A, lanes 2 and 4) than

DMSO (Fig. 17A, lanes 1 and 3) (densitometric units = 0.7 for both STS-treated

HEK293 and JHL15 and 1.0 and 1.1 for DMSO- treated HEK293 and JHL15 cells,

respectively) while its levels were similar in STS-treated (Fig. 17A, lane 5) and untreated

(Fig. 17A, lane 6) JHLa1 cells (densitometric units= 1.2 for both treatments). Conversely,

the mean % cells positive for the caspase-3 cleavage product (caspase-3p20) was

significantly ($p < 0.01$ by Student t test) higher in STS treated HEK293 (40.5 ± 0.9 %)

and JHL15 (36.4 ± 7.6 %) than in STS-treated JHLa1 (9.3 ± 0.1 %) cells, as determined

by immunocytochemistry with antibody specific for this product (Fig. 17B). The data

indicate that ICP10 PK interferes with caspase-3p32 cleavage/activation.

Because PARP cleavage by caspase-3 is regarded as a hallmark of caspase-3

dependent apoptosis (Tewari *et al.*, 1995), the second series of experiments was designed

to determine whether ICP10 blocks PARP cleavage. A 116 kDa band consistent with the uncleaved PARP was seen in extracts of mock- and STS-treated cells immunoblotted with PARP antibody (Fig. 17 C, lanes 1 to 6). By contrast, the 85 kDa band that is consistent with the PARP cleavage product was only seen in STS-treated HEK293 (Fig. 17 C, lane 2) and JHL15 (Fig. 17 C, lane 4) cells. It was not seen in STS-treated JHLa1 cells (Fig. 17 C, lane 5) nor in mock-treated HEK293, JHL15 or JHLa1 cells (Fig. 17 C, lanes 1,3,5). These data support the interpretation that ICP10 PK blocks caspase-3-dependent apoptosis.

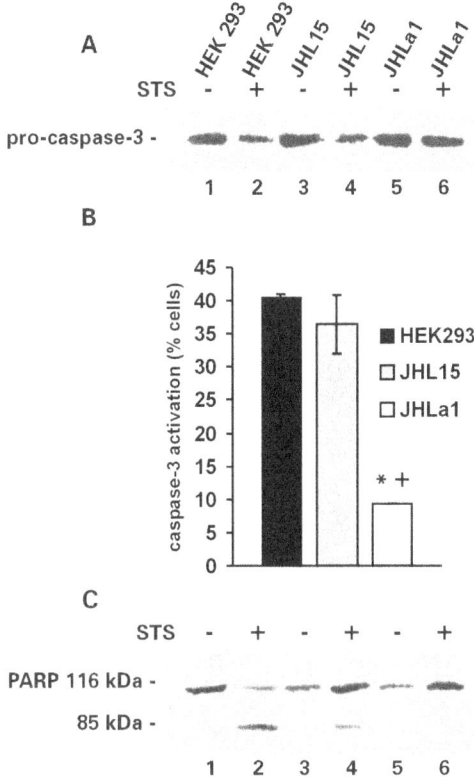

Fig. 17.   ICP10 PK inhibits caspase activation. (A) Proteins from extracts of

HEK293, JHL15 and JHLa1 cells, mock- (lanes 1, 3 and 5) or STS treated (250 nM, 24

hrs) (lanes 2, 4, 6) were resolved by SDS-PAGE (7 % acrylamide gels), transferred to

nitrocellulose membranes and immunoblotted with caspase3p32 antibody. **(B)** HEK293,

JHL15 and JHLa1 cells treated with 250 nM STS for 24 hrs were stained with

caspase3p20 antibody. Cells were counted in 5 randomly chosen microscopic fields and

the results are expressed as mean % positive cells $\pm$ SEM (* = $p < 0.01$ vs. HEK293 ; + =

$p < 0.01$ vs. JHL15 by Student t test). **(C)** The blot from panel A was stripped and

immunoblotted with anti-PARP antibody. An 85 kDa band consistent with the PARP

cleavage product was seen in STS-treated HEK293 (lane 2) and JHL15 (lane 4) but not in JHLa1 (lane 6) cells. Lanes are identified by numbers at the bottom of the panels.

The goal of the next series of experiments was to determine whether ICP10 PK has anti-apoptotic activity in the context of HSV-2 infection, and whether this activity is cell-type specific. Various cell lines [Vero, HEp-2, NIH3T3, PC12 (proliferating or neuronally-differentiated)] or primary neuronal cultures (cortical and hippocampal) were infected with HSV-2 or an HSV-2 mutant deleted in the PK domain of ICP10 (designated ICP10ΔPK), or mock-infected. In some experiments, HSV-1 and an HSV-2 mutant deleted in the RR domain of ICP10 (designated ICP10ΔRR) were used as controls.

4.   ICP10 PK does not modulate apoptosis in Vero or HEp-2-infected cells. HSV-2 was previously reported to induce (Koyama *et al.*, 1998) or inhibit (Sieg *et al.*, 1996) apoptosis in a cell type specific manner, but the identity of pro-apoptotic genes involved was not clarified. To determine whether ICP10 PK has a role in suppression of apoptosis in Vero cells, previously reported in this cell type for HSV-2 (Koyama *et al.*, 1998), they were infected with HSV-2, ICP10ΔPK (at m.o.i. 10), or mock-infected with growth medium and analyzed for apoptosis at 0.5, 4 and 24 hrs p.i. by TUNEL using fluorescent microscopy. Apoptosis was not seen in virus- or mock-infected cells at any time (data not shown). By contrast, both HSV-2 and ICP10ΔPK induced apoptosis (20-35 % apoptotic cells) in HEp-2 cells [as previously reported for HSV-2 (Koyama *et al.*, 1998)], suggesting that ICP10 PK (or other viral genes it regulates) is not responsible for the modulation of apoptosis (either suppression or induction) in these cell types.

5. **ICP10 PK is pro-apoptotic in mouse fibroblast cell line NIH3T3.** To determine

whether virus infection induces apoptosis in NIH3T3 cells, they were infected with HSV-

2, ICP10ΔPK (m.o.i. 10), or mock-infected with growth medium and analyzed for

apoptosis at 0.5, 4 and 24 hrs p.i. by TUNEL, using fluorescent and/or light microscopy.

Apoptosis was not seen at 0.5 hrs p.i. in mock or virus infected cells but by 4 hrs p.i.

HSV-2-infected cells began to show apoptotic degeneration (data not shown). At 24 hrs

p.i. apoptosis was still seen in HSV-2 infected cells (Fig. 18 B, D) but not in mock- (Fig.

18 A, D) or ICP10ΔPK (Fig. 18 C, D)-infected cells. By counting the apoptotic and non-

apoptotic cells (as described in Materials and Methods), it was determined that apoptosis

is induced at a significant level ($p < 0.01$, by ANOVA) in HSV-2 (43.4 ± 1.6 %) versus

mock (8 ± 2.6 %) or ICP10ΔPK (16.3 ± 4.9 %) infected cells (Fig. 18 D). These data

suggest that ICP10 PK (or a viral gene it regulates) is responsible for induction of

apoptosis in NIH3T3 cells.

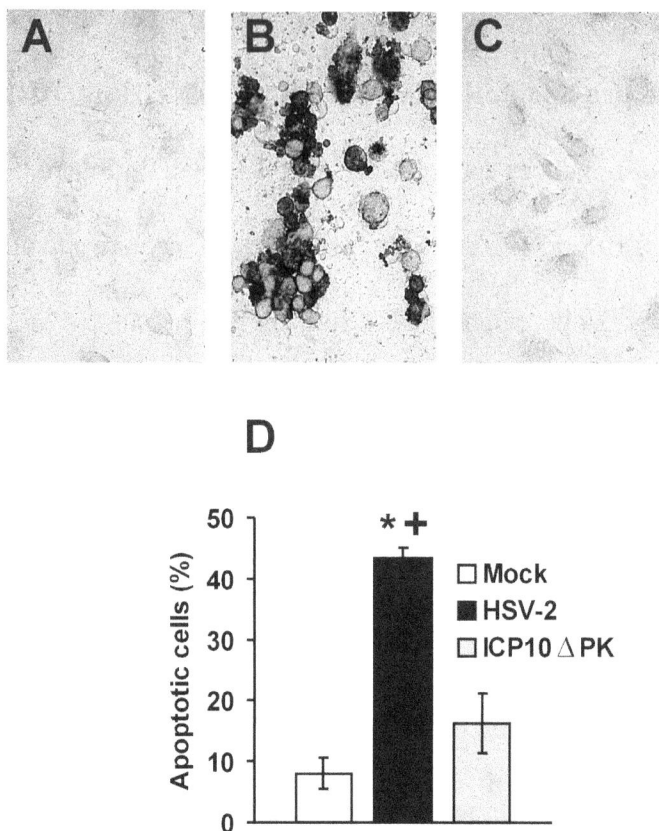

Fig. 18.   ICP10 PK induces apoptosis in NIH3T3 cells. Cells infected with HSV-2

(B), ICP10ΔPK **(C)** or mock-infected **(A)**, were assayed by TUNEL at 24 hrs p.i.

Apoptotic and non-apoptotic cells were counted and results were expressed as %

apoptotic cells ± SEM **(D)**. *, p<0.01 vs. mock; +, p<0.01 vs. ICP10ΔPK, by ANOVA.

6.   ICP10 PK inhibits virus-induced apoptosis in neuronally-differentiated

PC12 cells. Rat pheochromocytoma (PC12) cells proliferate in culture in the presence of

serum and display many characteristics of adrenal chromaffin cells. Within several days

of NGF exposure, these cells stop dividing and acquire numerous properties of mature

sympathetic neurons (e.g. neurite outgrowth, electrical excitability, and expression of

neuronal markers (Greene, 1978). To examine modulation of apoptosis in this system,

PC12 (proliferating or neuronally differentiated with NGF-for at least 12 days, as

described in Materials and Methods) were infected with HSV-2, ICP10ΔPK (m.o.i. 10) or

mock-infected with the appropriate growth medium and assayed by TUNEL at 24 hrs p.i.

Apoptotic and non-apoptotic cells were counted and results were expressed as %

apoptotic cells $\pm$ SEM. The percentages of apoptotic mock-infected cells were 2.3 $\pm$ 2.2

and 7.4 $\pm$ 3, for proliferating or neuronally differentiated cells, respectively, and they

were subtracted from each average obtained for virus-infected cells. There was no

statistical significant difference ($p > 0.05$, by ANOVA) among the percentages obtained

for mock (2.3 $\pm$ 2.2), HSV-2 (8.4 $\pm$ 1) or ICP10ΔPK (7.7 $\pm$ 1) infected proliferating PC12

cells (Fig. 19), suggesting that HSV-2 does not induce apoptosis and/or ICP10 PK is not

involved in modulation of apoptosis in these cells, similar to the previously described

Vero cells. By contrast, there was a significant difference ($p < 0.01$, by ANOVA) between

the percentage of apoptotic cells seen in neuronally differentiated PC12 cultures infected

with HSV-2 (20.8 $\pm$ 2.2) or ICP10ΔPK (65.6 $\pm$ 1.1) suggesting that ICP10 PK inhibits

specifically neuronal apoptosis triggered by virus infection.

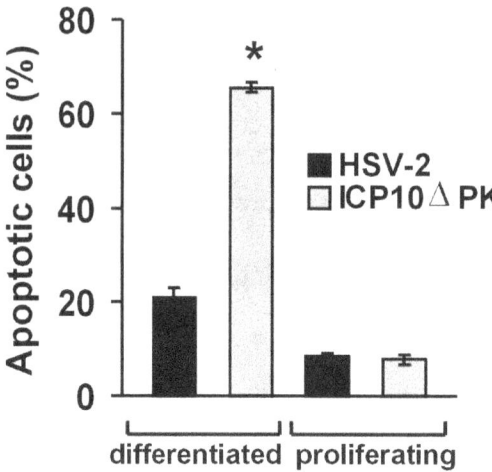

Fig. 19. **ICP10 PK inhibits virus-induced apoptosis in neuronally-differentiated**

PC12 cells. Proliferating or neuronally differentiated (NGF-dependent) PC12 cells (as

described in the text) were infected with HSV-2, ICP10ΔPK (m.o.i. 10) or mock-infected

with appropriate growth medium, and analyzed by counting TUNEL-positive and

TUNEL-negative cells at 24 hrs p.i. Results are expressed as % apoptotic cells ± SEM.

The percentage of apoptotic cells from mock-infected cultures was subtracted from each

average obtained for viruses. *, p<0.01 vs HSV-2, by ANOVA.

7.   HSV-2 requires the ICP10 PK gene and activation of Ras/Raf/MEK/ERK pathway for efficient virus replication in PC12 cells. HSV may establish latency in neurons (Saldanha *et al.*, 1986; Lynas *et al.*, 1993; Steiner and Kennedy, 1991) and the ability of the virus to establish a persistent infection may help evade the triggers that activate the cellular death program (Koyama *et al.*, 2000). Therefore, it seemed of interest to determine whether the distinct effect on apoptosis of HSV-2 and ICP10ΔPK is due to (or implicated in) a different pattern of virus replication. Proliferating (Fig. 20 A) or neuronally differentiated (Fig. 20 B) PC12 cells were infected with HSV-2 or ICP10ΔPK (m.o.i. 5). Cells and supernatants were collected at various times p.i. and viral titers determined by plaque assay as described (Aurelian, 2000). Fig. 20 A, B shows that HSV-2 and ICP10ΔPK have replication kinetics (in both proliferating or neuronally differentiated PC12 cultures) similar to those reported in other cell types with the ICP10ΔPK mutant exhibiting lower titers (Smith *et al.*, 1998), suggesting that ICP10 PK is required for efficient virus growth. However, since both viruses grow in PC12 cells (regardless of the differentiation state of the cells) their replication cannot be correlated with apoptosis modulation. The Ras/Raf/MEK/ERK pathway was also reported to be required for efficient HSV-2 replication in certain cell types (Smith *et al.*, 2000). To examine whether this holds true also in the context of virus infection of PC12 cells, cultures [proliferating (Fig. 20 C or neuronally differentiated (Fig. 20 D)] were pre-treated with the specific MEK inhibitor U0126 (10μM) or the specific PI3-K inhibitor LY294002 (50μM) for 1 hr and infected with HSV-2 (m.o.i. 5) in the presence of inhibitors. At various times p.i. cells and supernatants were collected and viral titers determined by plaque assay as described (Aurelian, 2000). As seen in Fig. 20 C. D., inhibition of the PI3-K pathway did not affect virus replication kinetics in PC12 cells regardless of their differentiation state. By contrast, MEK inhibition resulted in a significant reduction in viral titers, suggesting that the Ras/Raf/MEK/ERK pathway is required for HSV-2 growth in proliferating or neuronally differentiated PC12 cells. The effect of inhibition of Raf/MEK/ERK and PI3-K/Akt pathways on PC12 cell survival was not studied.

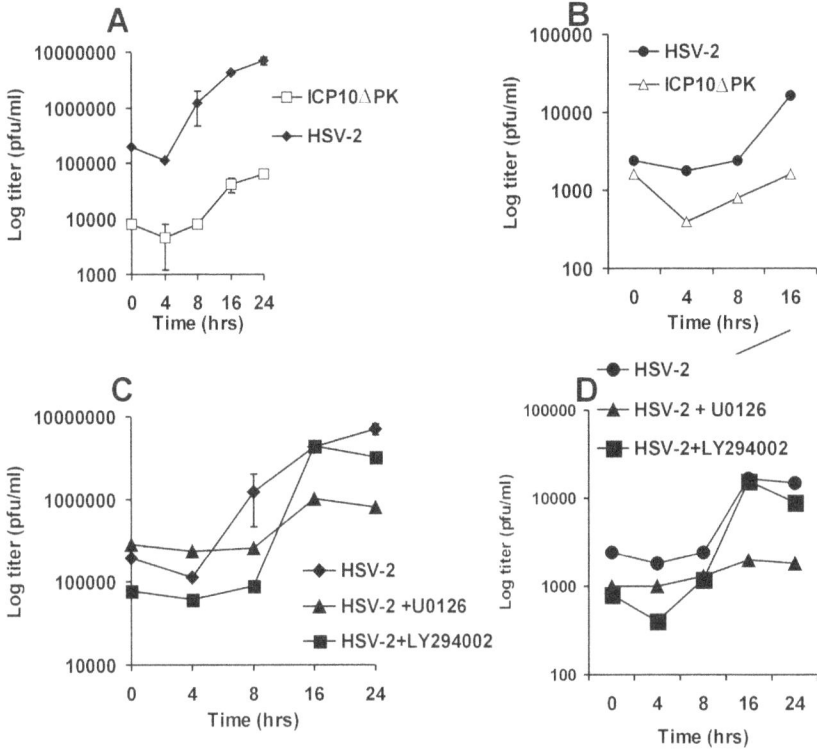

Fig. 20. **HSV-2 requires the ICP10 PK gene and activation of Ras/Raf/MEK/ERK pathway for efficient virus replication in PC12 cells.** Proliferating **(A)** or neuronally differentiated **(B)** PC12 cells (as described in Materials and Methods) were infected with HSV-2 or ICP10ΔPK (m.o.i. 5) and viral titers were determined at various times p.i. by plaque assay. Proliferating **(C)** or neuronally differentiated **(D)** PC12 cells were pretreated with U0126 (10 μM) or LY294002 (50 μM) for 1 hr and infected with HSV-2 (m.o.i. 5) in the presence of inhibitors. Viral titers were determined at various times p.i. by plaque assay.

8. **ICP10 PK protects CNS neurons from virus-induced apoptosis.** In order to determine whether ICP10 PK modulates the cell death of CNS neurons, primary cultures of cortical or hippocampal neurons were infected with HSV-2, ICP10ΔPK, ICP10ΔRR, HSV-2 (R), HSV-1 (m.o.i. 10) or mock-infected with growth medium, and examined by TUNEL at 24 hrs p.i. The proportion of apoptotic cortical neurons (Fig. 21A) was significantly higher in cultures infected with HSV-1 than HSV-2 (43.5 ± 3.6 % and 18.5 ± 2 %, respectively, p<0.01 by ANOVA). The increased resistance to apoptosis of HSV-2-infected cultures was due to ICP10 PK because high levels of apoptosis were also observed in cultures infected with ICP10ΔPK but not ICP10ΔRR- which retains the ICP10 PK DNA (48.2 ± 2.5 and 22.2 ± 1.2 %, respectively, p<0.01 by ANOVA). Primary cultures of hippocampal neurons were infected as above and analyzed by TUNEL at 16 and 24 hrs p.i. The data shown in Fig. 21 B represent the results of three independent experiments using 24 hrs infected cells. The % apoptotic cells was significantly lower in cultures infected with HSV-2 (5.7 ± 0.5 %) or the revertant virus HSV-2 (R) (7.9 ±1 %) than ICP10ΔPK (68 ± 3.9 %); p<0.01, by ANOVA). Apoptosis was also significantly lower in cultures infected with ICP10ΔRR than ICP10ΔPK (23 ± 2.1 % and 68 ± 3.9 %, respectively; p<0.01, by Student t test). Apoptosis was induced in a time-dependent fashion with a maximal increase occurring around 16 hrs p.i. (% apoptotic cells of 44 ± 3.5 and 5.4 ± 1.2 for ICP10ΔPK and HSV-2 respectively). The proportion of TUNEL-positive hippocampal neurons (Fig. 21 B) was also significantly higher in cultures infected with HSV-1 than HSV-2 (50.4 ± 4.5 % and 5.7 ± 0.5 %, respectively, p<0.01 by ANOVA), suggesting that the HSV anti-apoptotic activity is serotype-specific: Hoechst staining (shown for a representative nucleus) revealed the presence of nuclear fragmentation characteristic of apoptosis in ICP10ΔPK (Fig. 21 C), and HSV-1 (Fig. 21 D) but not in HSV-2 (Fig. 21 E) infected cells. The data indicate that the HSV anti-apoptotic activity is serotype-specific since HSV-1 induces significantly higher levels of apoptotic cell death than HSV-2- whose anti-apoptotic activity is due to ICP10 PK.

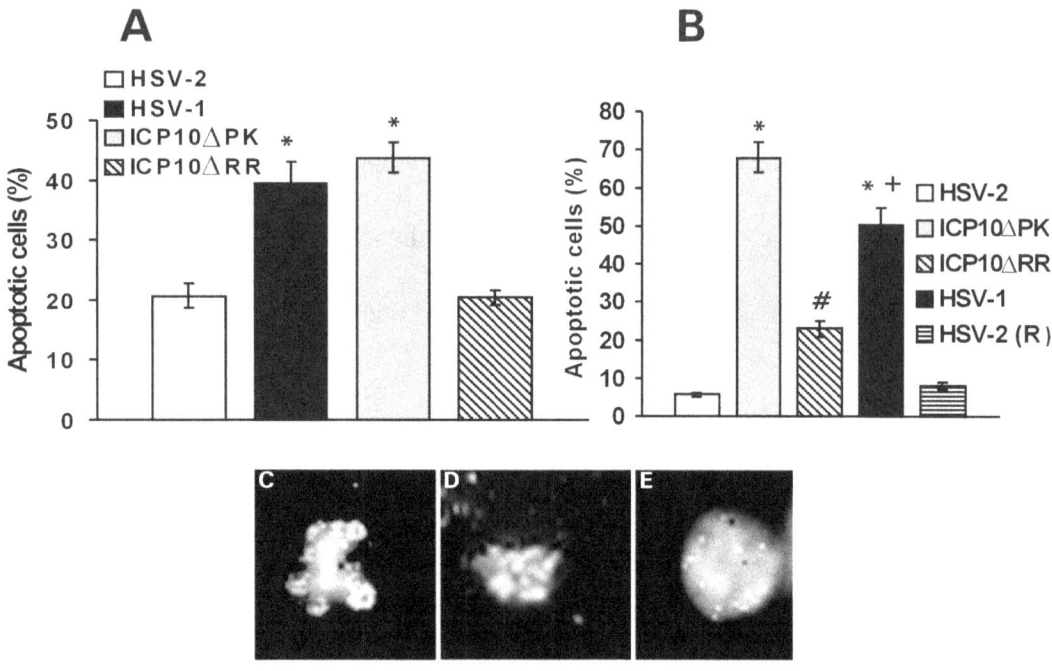

Fig. 21. ICP10 PK protects CNS neurons from virus-induced apoptosis. Primary

cortical **(A)** or hippocampal **(B)** cultures were infected with HSV-2, HSV-2 (R),

ICP10ΔPK, ICP10ΔRR or HSV-1 (m.o.i. 10) and analyzed by TUNEL at 24 hrs p.i.

Apoptotic (TUNEL- positive) and non-apoptotic (TUNEL-negative) cells were counted in

5 randomly chosen microscopic fields and the results are expressed as mean % apoptotic

cells ± SEM. The % apoptotic cells averages from mock-infected cultures were subtracted.

*, p < 0.01 vs. HSV-2 or HSV-2 (R); #, p < 0.05 vs HSV-2 or HSV-2 (R); +, p > 0.05 vs.

ICP10ΔPK, by ANOVA. **(C, D, E)** Hoechst staining of representative nuclei from cells in

panel B [ICP10ΔPK **(C)**, HSV-1 **(D)** or HSV-2 **(E)**].

9.   ICP10 PK inhibits caspase activation and PARP cleavage in primary cultures of hippocampal neurons. To examine the mechanism by which HSV, and in particular ICP10 PK, modulates apoptosis in hippocampal neurons, caspase activation was analyzed in infected primary cultures using Promega's CaspACE FITC-VAD-FMK *In Situ* Marker. Cultures were infected with HSV-2, ICP10ΔPK, ICP10ΔRR, or HSV-1 (m.o.i. 10) or mock-infected for 24 hrs before incubation with the FITC-conjugated cell permeable irreversible pan caspase inhibitor VAD-FMK [specifically binds to the active sites of caspases 3, 6 and 7 (Bossy-Wetzel *et al.*, 1998; Schotte *et al.*, 1999), and examined by fluorescence microscopy. No signal was detected in mock (not shown) or HSV-2-infected cultures (Fig. 22 A) and only low levels were detected in ICP10ΔRR-infected cultures (Fig. 22 B). By contrast, ICP10ΔPK- and HSV-1- infected cultures (Fig. 22 C and D, respectively) showed increased binding of the caspase inhibitor as evidenced by FITC signal in approximately 60-70 % of cells, supporting the conclusion that ICP10ΔPK and HSV-1 trigger caspase-dependent apoptosis in hippocampal neurons in culture. Similar results were obtained by counting cells stained with antibody specific for activated caspase-3 [recognizes the 17-20 kDa fragment of the active caspase (Fernandez-Alnemri *et al.*, 1994; Nicholson *et al.*, 1995; Hu *et al.*, 2000). As shown in Fig. 22 E, the % caspase3p20-positive cells was significantly ($p<0.05$ by ANOVA) higher in cultures infected with ICP10ΔPK (52.6 ± 9.6 %) or HSV-1 (47.4 ± 7.4 %) than HSV-2 (19.3 ± 2 %), HSV-2 (R) (17.5 ± 2.3 %) or ICP10ΔRR (26 ± 0.2 %), suggesting that caspase-3 is implicated in ICP10ΔPK and HSV-1- induced apoptosis in these cells. Furthermore, these data also evidence a role for ICP10 PK in inhibition of caspases (in particular caspase-3) activation. Because PARP cleavage by caspase-3 is a hallmark of the commitment to undergo apoptosis (Tewari *et al.*, 1995), it was next determined whether HSV infection leads to PARP cleavage. Hippocampal cultures were infected with HSV-1, HSV-2, ICP10ΔPK or ICP10ΔRR (m.o.i. 10) or mock-infected for 24 hrs and stained with antibody specific for the p85 (cleaved) fragment of PARP (Cookson *et al.*, 1999; Perng *et al.*, 2000; Thomson *et al.*, 2000). A very low percentage of cells (4.2 %) stained in the mock-infected cultures, and this % was subtracted from each average obtained for the virus infected cultures. The % p85PARP-positive cells was also very low in HSV-2 (5.7 ± 0.2) and ICP10ΔRR (3.2 ± 1) (p>0.05 by ANOVA), similar to that in mock-infected cultures (Fig. 22 F). By contrast, 29.6 ± 2.2 % of HSV-1- and 35.3 ± 2.5 % of ICP10ΔPK-infected cells were positive for p85 PARP (p<0.001 vs mock , ICP10ΔRR or HSV-2, by ANOVA). Staining was seen both in the nucleus and the cytoplasm (data not shown), as previously reported for p85 PARP (Cookson *et al.*, 1999; Perng *et al.*, 2000; Thomson *et al.*, 2000). The data are comparable to those obtained by TUNEL, and indicate that ICP10 PK protects hippocampal cultures from virus-induced apoptosis by inhibiting caspase-3 activation and PARP cleavage.

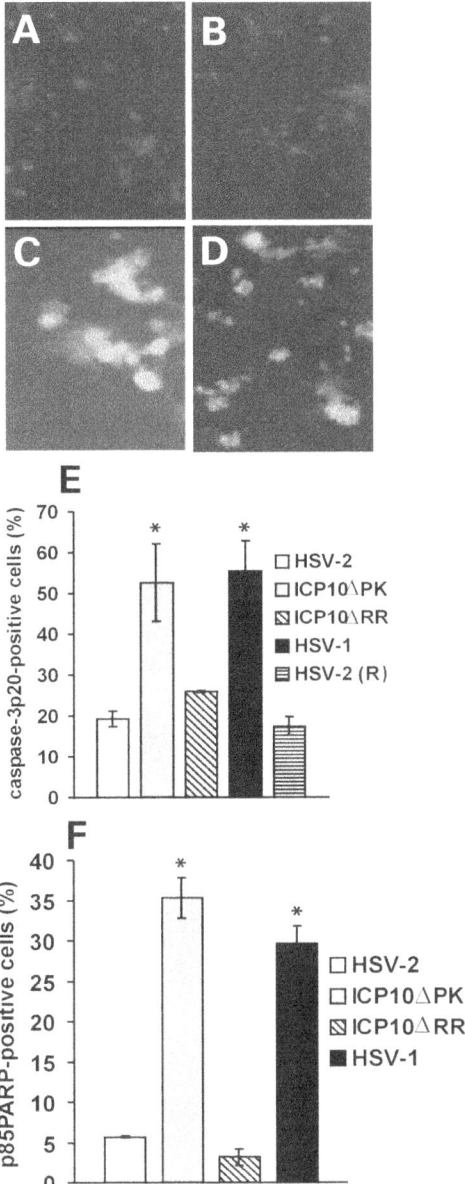

Fig. 22. ICP10 PK inhibits caspase activation and PARP cleavage in primary cultures of hippocampal neurons. Primary cultures of hippocampal neurons were mock-infected or infected with HSV-2 **(A)**, ICP10ΔRR **(B)**, ICP10ΔPK **(C)**, or HSV-1 **(D)** (m.o.i. 10), treated with the FITC conjugated-pan caspase inhibitor VAD-FMK for 20 min at 24 hrs p.i., and analyzed by fluorescence microscopy. **(E)** Primary cultures of

hippocampal neurons infected (24 hrs) with HSV-2, HSV-2 (R), ICP10ΔPK, ICP10ΔRR or HSV-1 (m.o.i 10) were stained with antibody specific for the p20 kDa fragment of active caspase-3 by immunoperoxidase method. Cells were counted and results were expressed as mean % caspase-3p20-positive cells ± SEM. The number of caspase-3p20-positive cells from mock-infected cultures was subtracted from each average. *, $p < 0.05$ vs HSV-2, HSV-2 (R) or ICP10ΔRR by ANOVA. **(F)** Primary cultures of hippocampal neurons infected as in (A) were stained with antibody specific for the p85 kDa fragment of PARP by the immunoperoxidase method. Cells were counted and results were expressed as mean % p85PARP-positive cells ± SEM. The number of p85PARP-positive cells from mock-infected cultures was subtracted from each average. *, $p < 0.001$ vs HSV-2 or ICP10ΔRR by ANOVA.

10.   Apoptotic cells in virus-infected-hippocampal cultures are neurons. Primary hippocampal cultures are likely to consist of various cell subpopulations. Therefore, it seemed essential to identify the cell type(s) present in these cultures and/or that become apoptotic upon infection with ICP10ΔPK or HSV-1. In the first series of experiments, hippocampal cultures were stained with PE-labeled NF-160 and TuJ1 (which are specific for postmitotic neurons), or GFAP and GalC antibodies (which are respectively specific for astrocytes and oligodendrocytes). No GalC immunoreactivity was detected in these cultures (data not shown), consistent with previous reports that such cultures are free of oligodendrocytes (Bertollini *et al.*, 1997). About 85-90 % of cells were identified as neurons (based on NF-160 or TuJ1 staining) and about 5-8 % as astrocytes (based on GFAP staining). In the second series of experiments, cell cultures were infected for 24 hrs with ICP10ΔPK, HSV-1 or HSV-2 (m.o.i. 10), or mock-infected, and double stained with FITC-labeled dUTP (TUNEL) and PE-labeled antibodies specific for neurons (NF-160 and TUJ1) or astrocytes (GFAP) as described in Materials and Methods. HSV-2-infected cultures stained with NF-160 antibody (Fig. 23 A) but they were mostly TUNEL-negative (Fig. 23 B) . Mock-infected cultures also stained with these antibodies (Fig. 24 C) and were TUNEL-negative (Fig. 23 D). By contrast, TUNEL-positive cells were seen in cultures infected with ICP10ΔPK (Fig. 23 E), and these cells stained with TuJ1 (Fig. 23 F, G). As previously described (Brazelton et al., 2000), the TuJ1 staining (PE) localized in the cell bodies and projections (Fig. 23 F), while the FITC staining (TUNEL) was primarily nuclear (Fig. 23 G). Some cells showed TUNEL staining in the cytoplasm (Fig. 23 E, G), presumably representing leakage of DNA fragments from the nucleus in late stage apoptotic cells (LaFerla and Gilbert, 1997). GFAP-stained cells in HSV-2-infected cultures (Fig. 23 I) were TUNEL-negative (Fig. 23 H, J). Double-labeling TUNEL-NF-160 studies were complicated by the fact that there was a decrease in NF-160 immunoreactivity in the vast majority of cells (data not shown). Downregulation of NF levels due to various treatments was previously described in neuronal systems (Beitner-Johnson *et al.*, 1992; Bunnemann *et al.*, 2000) and it may also constitute a characteristic feature of virus-induced apoptosis in these cells. Co-localization of TUNEL with TuJ1 staining was also seen for HSV-1-infected neurons (data not shown). The data indicate that

124

the apoptotic cells are neurons. By inference, ICP10 PK protects hippocampal neurons against virus-induced apoptosis.

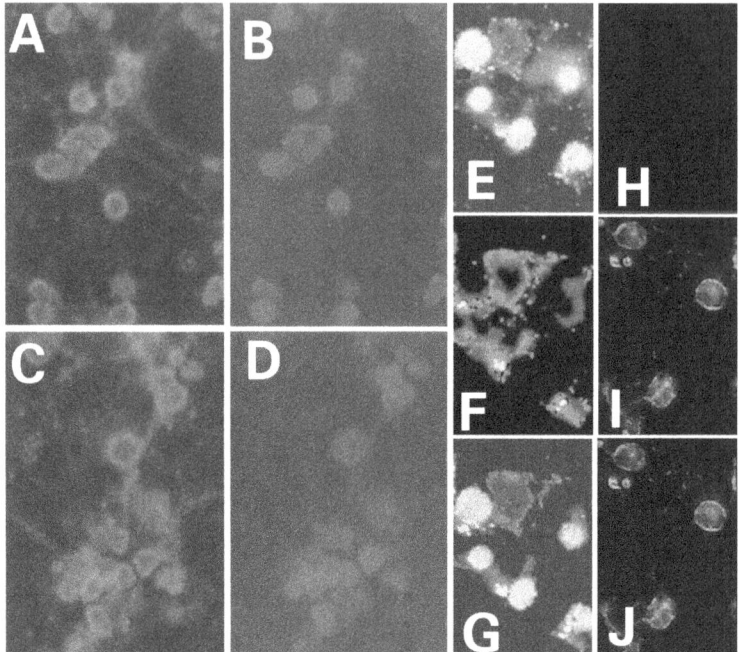

Fig. 23. TUNEL-positive cells in virus infected- hippocampal cultures are neurons.
Primary hippocampal cultures infected with HSV-2 (24 hrs; m.o.i. 10) **(A, B)** or mock-infected **(C. D.)** were stained with NF-160 antibody **(A, C)** and assayed by TUNEL **(B, D)**.Similar results were obtained with TuJ1 antibody (data not shown). ICP10ΔPK infected cultures were labeled by TUNEL **(E, H)** and stained with TUJ1 **(F)** or GFAP staining **(I)**. TUNEL co-localized with TuJ1 **(G)**, but not GFAP **(J)** staining.

11.   Apoptosis is not related to virus multiplication. Since HSV-1 and HSV-2-infection of the CNS correlates with a different neurological outcome, it seemed of interest to determine whether this effect is associated with a different replication pattern in CNS (i.e. hippocampal) neurons. Virus replication kinetics in primary cultures of hippocampal neurons was analyzed by single step growth assays. Primary cultures of hippocampal neurons were infected with HSV-2, ICP10ΔPK, ICP10ΔRR, or HSV-1 at m.o.i 5 and virus titers were determined at 0- 48 hrs p.i. by plaque assay as described (Aurelian, 2000). As shown in Fig. 24, HSV-2 and HSV-1 replicated well in these cells, with growth curves similar to those previously reported for other cell types (Smith *et al.*, 1994). Replication began at 3-5 hrs p.i. reaching maximal titers at 15-24 hrs p.i. ICP10ΔPK (apoptotic) and ICP10ΔRR (non-apoptotic) were growth defective in these cells. Similar patterns of virus replication were seen at other m.o.i. tested (1-10). These data suggest that the ICP10 gene is required for virus growth and that viral replication cannot be correlated with the ability of wt HSV or HSV-2 mutants to modulate apoptosis.

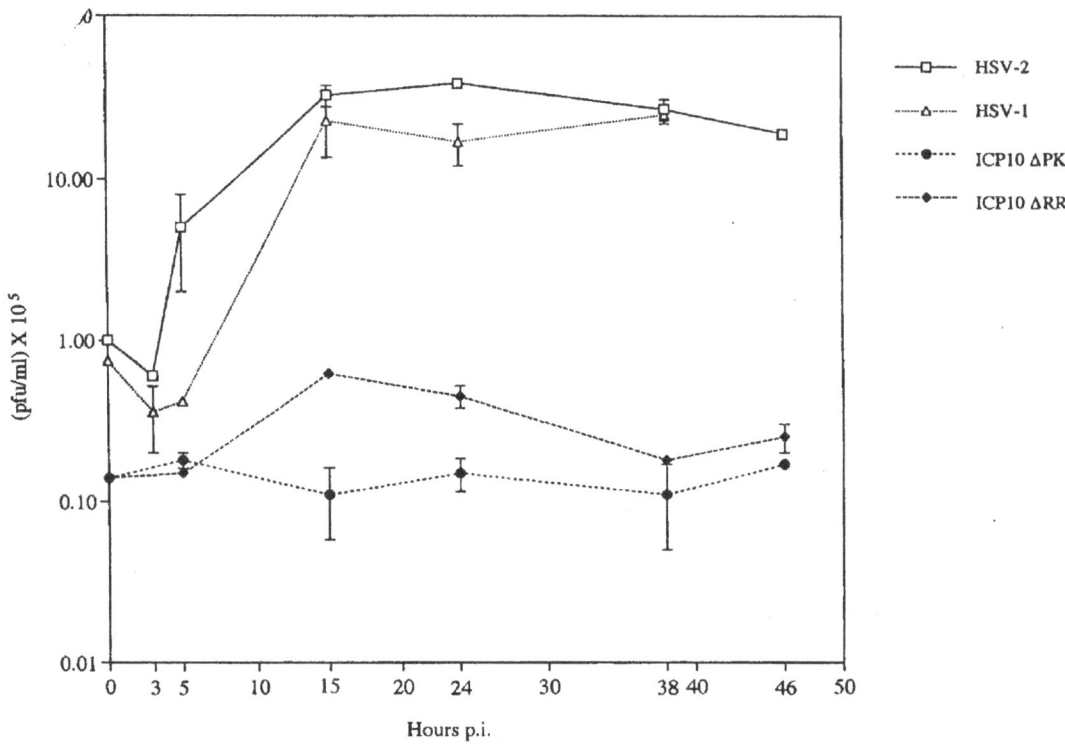

Fig. 24. Virus replication kinetics in primary cultures of hippocampal neurons.

Triplicates of primary cultures of hippocampal neurons were infected with HSV-2, HSV-1, ICP10ΔPK and ICP10ΔRR at 5 pfu/cell. Viral titers were determined at various times between 0 and 48 hrs p.i. by plaque assay. Results are expressed as mean pfu/ml ± SEM.

12.   Discussion- Specific Aim 1

Viruses have evolved various strategies to prevent apoptosis, including expression of Bcl-2 homologs, inhibition of caspases, and repression of p53 activity (reviewed by Hardwick *et al.*, 1998). HSV-1 and HSV-2 have anti-apoptotic activity that is cell type specific and has been attributed to the HSV-1 and HSV-2 gene US3, and to the HSV-1 genes ICP34.5, US5, ICP27 and LAT, which function by a still poorly understood mechanism (Chou and Roizman, 1992; Galvan and Roizman, 1998; Aubert and Blaho, 1999; Hata *et al.*, 1999; Jerome *et al.*, 1999; Perng *et al.*, 2000).

Previous studies from our laboratory showed that the HSV-2 gene ICP10 PK activates the Ras/Raf/MEK/ERK signaling pathway in non-neuronal cells (Smith *et al.*, 1994; Hunter *et al.*, 1995; Smith *et al.*, 2000). Because this pathway was involved in the control of apoptosis and functions as one of the main survival pathways in neurons (reviewed by Kaplan and Miller, 2000), we hypothesized that ICP10 PK may also be involved in modulation of apoptosis. The experiments presented in this section (i.e. Specific Aim 1) were designed to test this hypothesis. We found that the ICP10 PK gene modulates apoptosis in a cell-type specific manner. The outcome of this activity (as it refers to cell survival or apoptosis) correlates with previous reports which showed that activation of Ras/Raf/MEK/ERK pathway may lead to increased cell survival or apoptosis, depending on the cell type (reviewed by Downward, 1998). The following comments seem pertinent with respect to these findings.

Cell death can be due to apoptosis, necrosis, or a combination of both (Ankarcrona *et al.*, 1995), and the suitability of various assays for the detection of apoptosis has recently come under scrutiny (Chariaut-Marlangue and Ben-Ari, 1995; Cohen *et al.*, 1992). Our definition of apoptosis is based on a multiplicity of criteria (cell survival, nuclear morphology, TUNEL, DNA fragmentation, caspase-3 activation and PARP cleavage). According to these criteria, STS and D-Mann induced apoptosis in HEK293 and JHL15 cells, which express the PK-negative ICP10 mutant p139, but not in JHLa1 cells, which express ICP10 (Luo and Aurelian, 1992; Smith *et al.*, 1994). Protection requires a functional ICP10 PK activity, because: i) the JHLa1 cells and JHL15 cells were similarly established from HEK293 cells, and ii) p139 is expressed as well as ICP10 ((Luo and Aurelian, 1992; Smith *et al.*, 1994). The exact mechanism by which a functional ICP10 PK inhibits STS- or D-Mann-induced apoptosis is unclear. However, it is likely that the constitutive activation of Ras/Raf/MEK/ERK pathway mediated by ICP10 PK in JHLa1 (but not JHL15 or HEK293) (Luo and Aurelian, 1992; Smith *et al.*, 1994) is responsible for the protection against these specific apoptotic stimuli. Previous reports that showed a protective effect of Ras/Raf/MEK/ERK pathway activation against apoptosis induced by staurosporine (Deng *et al.*, 2000) support this interpretation. However, the contribution of other signaling pathways (e.g. PI3-K/Akt) cannot be excluded. The mechanism responsible involves ERK-mediated phosphorylation of Bcl-2 at Ser 70, a process which presumably increases the stability of Bcl-2-Bax heterodimers and promotes cell survival (Deng *et al.*, 2000). In contrast to STS [which induces apoptosis by inhibiting certain cellular kinases, such as PKC (Deng *et al.*, 2000)], D-Mann induces apoptosis by osmotic shock associated with oxidative stress (Malek *et al.*, 1998; Greene *et al.*, 1999) and this effect is inhibited by activated ERK (Guyton *et al.*, 1996).

The next series of experiments were designed to examine whether ICP10 PK has anti-apoptotic activity also in the context of virus infection. It also seemed of interest to determine whether ICP10 PK is involved in modulation of the previously reported cell type-specific anti- or pro-apoptotic activity of HSV-2 (Koyama *et al.*, 1998; Koyama *et al.*, 2000). To answer these questions, various cell lines or primary cultures were infected with wild type HSV-2 or an HSV-2 mutant deleted in the PK domain of ICP10 (designated ICP10ΔPK). In some experiments, the HSV-2 mutant, ICP10ΔRR (deleted in the RR domain of ICP10) and the revertant HSV-2 [HSV-2 (R)] were added as controls . The construction of the HSV-2 mutants and HSV-2 (R) was previously reported (Aurelian and Smith, 2000). To examine whether the HSV-1 homolog of ICP10 (designated ICP6PK) is also involved in modulation of apoptosis, in some experiments, cells were also infected with wild type HSV-1, or two HSV-1 mutants: one that has over 90 % of the ICP6 gene deleted (designated ICP6Δ), the other that has a deletion in the RR domain of ICP6 (designated *hr*R3). The construction of these mutants was also previously reported (Goldstein and Weller, 1987; Goldstein and Weller, 1988).

Previous reports (Koyama *et al.*, 1998; Koyama *et al.*, 2000) have shown that in Vero and HEp-2 cells, HSV-2-infection suppresses or triggers apoptosis, respectively, but the viral genes responsible for these activities were not identified. In the experiments described in this report, Vero and HEp-2 cells were mock-infected with growth medium, or infected with HSV-2 or the ICP10ΔPK mutant and apoptosis was determined by TUNEL at various times p.i. Both HSV-2 and ICP10ΔPK behaved similarly in these cells, in that none induced apoptosis in Vero, but both induced it at 24 hrs p.i. in 20-30 % of HEp-2 cells, suggesting that ICP10 PK does not modulate cell death in these cells. However, in NIH3T3 cells, apoptosis was induced by HSV-2, but not ICP10ΔPK, suggesting that ICP10 PK is pro-apoptotic. Interestingly, Ras itself has also been reported to have anti- as well as pro-apoptotic effects. Many cases in which Ras and its effectors (including Raf/MEK/ERK pathway) have been found to induce apoptosis involve either lymphocytes or fibroblasts. In fibroblasts, overexpression of an activated Ras oncogene promotes apoptosis, whereas v-ras promotes fibroblast apoptosis even in the absence of obvious stress (Fukasawa and Vande Woude, 1997). It is believed that, in fibroblasts, this pro-apoptotic effect of Ras is tied in with the cell-cycle regulation and is part of a protective response of cells against excessive activation of the Ras/Raf/MEK/ERK mitogenic pathway (Downward, 1998). Since ICP10 PK also has oncogenic potential due to its activation of Ras/Raf/MEK/ERK pathway in a variety of cells, including fibroblasts (Smith *et al.*, 1992; Smith *et al.*, 1994; Smith *et al.*, 2000), it is likely that the anti-apoptotic activity of ICP10 PK is modulated in the same fashion as described for activated Ras, and the experiments described here seem to support this conclusion.

In neuronal cells, the Ras/Raf/MEK/ERK pathway is involved in neuronal survival and function (Kaplan and Miller, 2000). To determine whether ICP10 PK is also involved in neuronal survival, we used PC12 cells and primary cultures of CNS neurons. PC12 cells proliferate in culture in the presence of serum and display many characteristics of adrenal chromaffin cells. Within several days of NGF exposure, these cells stop dividing and acquire numerous properties of mature sympathetic neurons (e.g. neurite outgrowth, electrical excitability, and expression of neuronal markers (Greene, 1978). Cells (either proliferating or neuronally differentiated with NGF, as described in Materials and Methods) were infected with HSV-2 or ICP10ΔPK, or mock-infected. Infection of proliferating PC12 cells with HSV-2 or ICP10ΔPK did not affect the cell survival. By contrast, a significant difference was found between the two viruses in their ability to modulate apoptosis in NGF-dependent, neuronally differentiated PC12 cells: HSV-2 induced low levels (about 20 %) whereas ICP10ΔPK induced high levels of apoptosis (in more than 60 % of cells), suggesting that ICP10 PK promotes the survival of virus-infected cells in this system. At this time it is not known whether ICP10 PK promotes the survival of differentiated PC12 cells by activating the Ras/MEK/ERK survival pathway. Our results show that ICP10 PK and activation of Ras/MEK/ERK (but not PI3-K/Akt) are required for efficient virus replication in PC12 cells (when they are proliferating or neuronally differentiated) but the contribution of these pathway to the survival of HSV-2-infected cells was not studied. However, it is unlikely that induction of apoptosis by the specific MEK inhibitor U0126 is responsible for the reduced titers of HSV-2, because previous studies showed that HSV-2 replicate well in apoptotic cells (Koyama *et al.,* 2000). These data correlate with previous reports that showed that ICP10 PK is required for viral regulatory IE gene expression and, consequently, for timely initiation of the protein cascade and HSV-2 growth (Smith *et al.*, 1998; Aurelian *et al.*, 1999).

To examine whether ICP10 PK has anti-apoptotic activity in CNS neurons, primary rat hippocampal or cortical cultures were infected with HSV-2, HSV-2 (R), ICP10ΔPK, ICP10ΔRR, or HSV-1 and apoptosis was determined at various times p.i. by

TUNEL, specific nuclear morphology, effector caspases (and caspase-3, in particular) activation, and PARP cleavage. Cultures were infected in medium containing 10 % serum, a condition which allows the replication of ICP10ΔPK and ICP10ΔRR in non-neuronal cells (Smith *et al.*, 1998; Aurelian and Smith, 2000). All viruses used in this study had similar titers, a condition that allows infection with similar volumes of viral suspension (also matched by adding the same volume of growth medium for mock-infection) in order to control for the potential contribution of growth factors in the serum. The single step growth curves were done at m.o.i. of 5 but similar patterns of virus replication were observed (data not shown) at other m.o.i. tested (between 1 and 10 pfu/cell). In order to increase the likelihood of detecting protein alterations associated with apoptosis, we chose the m.o.i. 10 for these experiments, to allow for optimal representation of survival- and apoptosis-induced activities.

Significantly, HSV-2 and HSV-1 replicated in hippocampal cultures with growth kinetics similar to those reported previously for non-neuronal cells (Smith *et al.*,1994; Smith *et al.*, 1998). However, both ICP10ΔPK and ICP10ΔRR were growth defective. This defect is not due to improper adsorption or penetration, because immunoblotting with ICP10 antibody revealed the presence of both p95 and p175 in extracts of cells infected with ICP10ΔPK or ICP10ΔRR, respectively, at 0.5 hrs p.i., but the proteins were not seen in extracts of cells similarly infected with antibody-neutralized viruses, which can adsorb to but do not penetrate the cells (data not shown). These findings are consistent with the previously described experiments using PC12 cells and others (Aurelian *et al.*, 1999) that showed that mutants deleted in the RR or PK domain of ICP10 are growth compromised in neuronal (and therefore non-dividing) cells.

Apparently unrelated to virus replication, high levels of apoptosis were seen in cultures infected with ICP10ΔPK or HSV-1, but not HSV-2 or ICP10ΔRR (that retains the PK DNA), suggesting the unique structural and functional properties of ICP10 PK (that differentiate it from its HSV-1 counterpart) are responsible for its anti-apoptotic activity in CNS cells in culture. Double immunostaining [TUNEL- neuron- (astrocyte or oligodendrocyte) specific antibody] of hippocampal cultures showed that the apoptotic cells were neurons (which are in 85-90 % proportion in these cultures). We conclude that the anti-apoptotic activity in CNS neurons depends on a functional ICP10 PK because: i) ICP10ΔPK does not have genetic alterations other than this deletion (Smith *et al.*, 1998); ii) US3, the only HSV-2 gene previously shown to have anti-apoptotic activity in non-neuronal cells (Hata *et al.*, 1999), is functional in ICP10ΔPK (data not shown), and iii) the revertant virus HSV-2 (R) had anti-apoptotic activity similar to that of HSV-2.

It is particularly interesting that the percentage of TUNEL-positive (apoptotic) cells were also higher in hippocampal cultures infected with ICP10ΔRR than with HSV-2. Although this percentage was significantly lower than that seen in ICP10ΔPK-infected cultures, the observation suggests that the ICP10 RR domain also contributes to the anti-apoptotic activity of HSV-2 in hippocampal neurons. The exact mechanism responsible for this contribution is still unclear. We assume that the ICP10 RR domain functions downstream of caspase-3 activation in hippocampal neurons because the proportion of cells positive for activated caspase-3 was similar for ICP10ΔRR and HSV-2. However, we cannot exclude the possible contribution of protein modification resulting from *LacZ* insertion. Interestingly, the contribution of the ICP10 RR domain to the anti-apoptotic activity of HSV-2 cannot be seen in cortical cultures, where HSV-2 and ICP10ΔRR trigger similar levels of apoptosis (albeit lower than HSV-1 or ICP10ΔPK). It is unclear at this time why the ICP10 RR domain would function differently in cortical versus hippocampal neurons. Cortical cultures also seem to be more resistant to virus-induced apoptosis because HSV-1 and ICP10ΔPK induced slightly lower levels of apoptosis than in hippocampal cultures. Interestingly, previous studies have shown that certain classes of neurons are more susceptible to HSV infection than others, but the mechanism involved is still unclear (Kristensson *et al.,* 1982). However, there is a wealth of information regarding the ability of HSV-1 to target hippocampal neurons in the case of human infection of the CNS (reviewed by Damasio and Van Hoesen, 1985). These characteristics may reflect the specific endogenous properties of hippocampal versus cortical neurons. Such properties may include a differential expression of HSV receptors, but the distribution/identity of these receptors in CNS tissue is not yet known (Schweighardt and Atwood, 2001). The hippocampus is a source of a relatively homogenous population of neurons, i.e. pyramidal neurons, which account for 85 % to 90 % of the total neuronal population (Goslin and Banker, 1998). A variety of interneurons have also been described in the hippocampus (Swanson *et al.*, 1987), but they are few in number compared to pyramidal cells. The existence of a single dominant cell type in the hippocampus (but not cortex), suggests that ICP10 PK has a more potent anti-apoptotic activity in pyramidal neurons than in the mixed population of cortical cells. However, the studies described in Specific Aim 2 were designed to examine the ICP10 PK anti-apoptotic activity in hippocampal neurons. Deciphering the specific features of cortical versus hippocampal neurons and their correlation with the ICP10 PK-mediated protection from virus-induced apoptosis, remains a challenge for the future. Notwithstanding, hippocampal neurons express NMDA receptors at particularly high levels, a property that was linked to the susceptibility of these neurons to anoxia (Collingridge and Bliss, 1987; Cotman *et al.*, 1988). Thus, experiments were designed (Specific Aim 3) to determine whether virus-induced apoptosis proceeds via the NMDA receptors.

Specific Aim 2. To investigate the targets of HSV-2-induced survival activity in hippocampal neurons [activation of Raf/MEK/ERK and/or PI3-K/Akt pathways, inhibition of pro-apoptotic JNK signaling pathway, and the respective transcription factors involved (CREB, fos, c-Jun, ATF-2, p53)] and to determine the effect of ICP10 PK on Bcl-2 family of proteins (Bcl-2, Bad) and their chaperones (e.g. Bag-1).

1.   ICP10 PK activates MEK/ERK in hippocampal neurons. The anti-apoptotic activity of ICP10 PK in hippocampal neurons raises the possibility that it is related to its ability to activate the MEK/ERK pathway (Smith *et al.*, 1994; Smith *et al.*, 2000). Cultures of hippocampal neurons were infected with HSV-2, ICP10ΔPK or HSV-1 (m.o.i. 10), or mock-infected, and analyzed for ERK activation by immunoblotting with antibodies specific for P-ERK1,2 and ERK1,2 as described in Materials and Methods. Cell extracts were studied at 0.5 and 24 hrs p.i. (0 hrs p.i. is at the end of adsorption). At 0.5 hrs p.i., P-ERK1,2 levels were significantly higher in HSV-2 infected than mock infected cells. The increased levels of P-ERK1,2 in HSV-2 infected cells reflect ERK activation by ICP10 PK because P-ERK1,2 levels were not increased in cells infected with ICP10ΔPK or HSV-1. ERK activation is a relatively rapid and transient event, since similar P-ERK1,2 levels were seen in all cultures at 24 hrs p.i. (Fig.25 A, B). The conclusion that ERK1,2 are activated within 30 min p.i. by HSV-2, but not ICP10ΔPK or HSV-1, is supported by densitometric scanning and data analysis as P-ERK/ERK ratios.

To examine the contribution of upstream components of the Ras survival pathway towards ERK activation, cultures of hippocampal neurons were infected with HSV-2 in the presence (or absence) of 20 μM of the MEK-specific inhibitor U0126 and cell extracts were examined for P-ERK1/2 by immunoblotting with specific antibody. P-ERK1/2 levels were significantly lower in cells treated with U0126 (Fig. 25 C, lane 3, Fig. 25 D) than in untreated cells (Fig. 25 C, lane 2; Fig. 25 D), suggesting that ERK activation is MEK dependent.

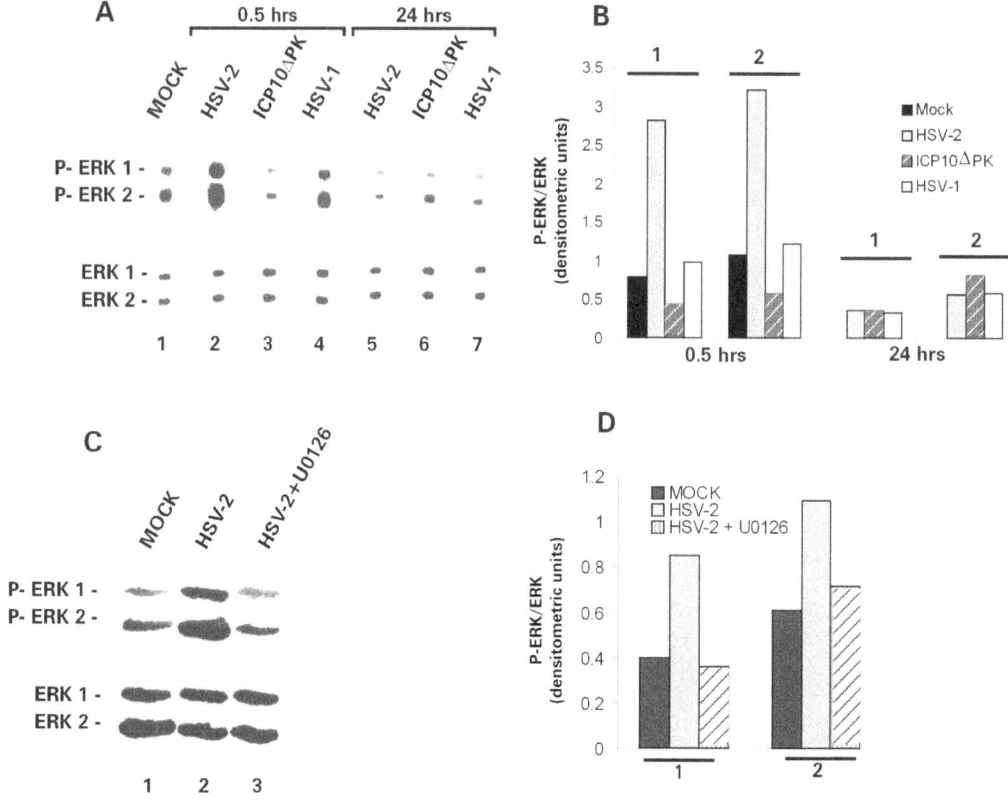

Fig. 25. **ICP10 PK activates MEK/ERK in hippocampal neurons. (A)** Hippocampal

neurons in culture were infected (10 pfu/cell) with HSV-2 (lanes 2 and 5), ICP10ΔPK

(lanes 3 and 6) or HSV-1 (lanes 4 and 7) or mock-infected (lane 1) and harvested at 0.5

and 24 hrs p.i. Proteins were resolved by SDS-PAGE (8.5% acrylamide gels), transferred

to nitrocellulose membranes and immunoblotted with antibody specific for P-ERK1,2

(upper bands). Blots were stripped and re-blotted with antibody to ERK1,2 (lower bands).

Lanes are identified by numbers at the bottom of the panel.**(B)** Densitometric scanning of

bands in panel A expressed as P-ERK/ERK ratios for P-ERK1 (1) and P-ERK2 (2). **(C)**

Extracts of hippocampal neurons infected with HSV-2 in the absence (lane 2) or presence (lane 3) of 20 µM U0126 or mock-infected (lane 1) were immunoblotted with antibody specific for P-ERK1,2 (upper bands) or ERK1,2 (bottom bands). Lanes are identified by numbers at the bottom of the panel. **(D)** Densitometric scanning of bands in panel C expressed as P-ERK/ERK ratios for P-ERK1 (1) and P-ERK2 (2).

2.   The anti-apoptotic activity of ICP10 PK requires activation of the Raf/MEK/ERK survival pathway. To examine whether activation of MEK/ERK and its upstream effector c-Raf by HSV-2 (i.e. ICP10 PK), two series of experiments were done. In the first series of experiments, hippocampal cells were mock- or HSV-2-infected in the absence or presence of 10 or 20 µM of the specific MEK inhibitor U0126 and examined by TUNEL at 24 hrs p.i. A dose-dependent increase in the % TUNEL-positive (apoptotic) cells was seen in HSV-2-infected cultures treated with U0126 (27 ± 1.6 % and 48.3 ± 2.8 % for 10 and 20 µM respectively) as compared to 10 ± 0.7 % in similarly infected but untreated cells (p<0.01, by ANOVA), suggesting that the anti-apoptotic activity of HSV-2 requires MEK/ERK activation (Fig. 26 A). Significantly, however, U0126 treatment had no effect on the survival of mock-infected cultures (Fig. 26 A), although U0126 inhibits MEK/ERK activation. The data suggest that MEK/ERK activation is not required for the basal survival of uninfected hippocampal neurons, at least under these experimental conditions. In the second series of experiments, we investigated the role of c-Raf kinase in HSV-2 anti-apoptotic activity by using a c-Raf dominant negative expression vector K375M (Dent *et al.*, 1995; Soh *et al.*, 1999). Hippocampal cultures were transfected with the dominant negative c-Raf K375M-FLAG expression vector for 24 hrs and infected with HSV-2 or mock infected for 24 hrs. 36 ± 2.2 % and 32.7 ± 3.6 % of hippocampal cells from mock or HSV-2 infected and transfected cultures, respectively, stained with antibody specific for the FLAG marker (Fig. 26 C and 26 D, respectively). Staining was minimal in non-transfected cultures (4.1 ± 0.3 %) (Fig. 26 B). The anti-apoptotic activity of HSV-2 was c-Raf-dependent as determined by the occurrence of increased levels of apoptosis at 24 hrs p.i. in HSV-2- but not mock- infected and transfected cultures (60.7 ± 5.7 % and 14.5 ± 3.5 %, respectively, $p<0.01$, by ANOVA) (Fig. 26 A). These data suggest that HSV-2 (and thereby ICP10 PK) requires the Raf/MEK/ERK survival pathway to exert its anti-apoptotic activity.

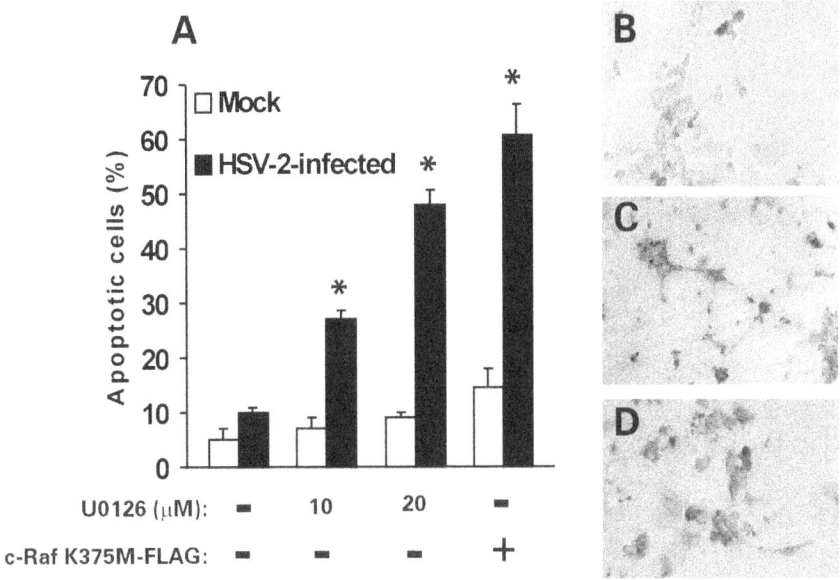

Fig. 26. **The anti-apoptotic activity of ICP10 PK requires activation of the Raf/MEK/ERK survival pathway. (A)** In the first paradigm, hippocampal cultures were infected with HSV-2 (m.o.i.10) or mock-infected, in the absence(-) or presence of U0126 (10-20 μM). In the second paradigm, hippocampal cultures were transfected with the dominant negative mutant c-Raf K375M-FLAG (+) and infected with HSV-2 (m.o.i 10) or mock-infected at 24 hrs post-transfection. All cultures were analyzed by TUNEL at 24 hrs p.i. (= 24 hrs U0126 treatment = 48 hrs post-transfection). Results shown are average of three independent experiments. *, p<0.01 vs mock by ANOVA. **(B)** Immunoperoxidase staining for FLAG at 48 hrs post-transfection (24 hrs p.i.) of non-transfected mock-infected hippocampal cultures, **(C)** cultures transfected with c-Raf-1 K375M-FLAG and mock-infected or **(D)** transfected with c-Raf-1 K375M-FLAG and HSV-2-infected.

3.   MEK/ERK activation does not require *de novo* viral protein synthesis. The

finding that MEK/ERK are not activated by ICP10ΔPK which does not replicate in

hippocampal cultures, suggests that activation may require *de novo* viral protein synthesis.

However, activation is seen at 30 min p.i., before onset of viral protein synthesis (2-3 hrs

p.i.). To examine whether viral protein synthesis is required for MEK/ERK activation,

hippocampal cultures were infected with HSV-2 or UV-inactivated HSV-2 (m.o.i. 10),

which penetrates the cells but is defective in protein synthesis (Purifoy and Powell, 1977).

Infection was in the absence or presence of U0126 (20 μM). Cell extracts collected at 30

min p.i. were examined for ERK activation by immunoblotting with P-ERK1,2 antibody

and ERK1,2 antibody (control). P-ERK1,2 levels were increased in cultures infected with

HSV-2 (Fig. 27 A, lane 3) or UV-inactivated HSV-2 (Fig. 27 A, lane 5) relative to mock-

infected cells (Fig. 27 A, lane 1) and the increase was virtually identical for both HSV-2

and UV-inactivated HSV-2. P-ERK1,2 levels were not increased when infection was in the

presence of U0126 (Fig. 27 A, lanes 4, 6) and ERK1,2 levels were similar in all samples.

Densitometric scanning and analysis of P-ERK/ERK ratios (Fig. 27 B) confirmed the

conclusion that ERK is activated equally well by HSV-2 and UV-inactivated HSV-2,

suggesting that MEK/ERK activation is independent of *de novo* viral protein synthesis.

This conclusion carries the implication that pathway activation is due to the ICP10

protein present in the tegument of HSV-2 virion and released into the cytoplasm of the

infected cell upon virus uncoating. Thus, it may also be implied that ICP10ΔPK causes

apoptosis because its mutant ICP10 protein [p95, which is also incorporated into the virion

tegument (Smith *et al.*, 1997)] is not released into the cytoplasm. To test this interpretation, hippocampal cultures were infected with HSV-2, ICP10ΔPK, ICP10ΔRR (expresses the mutant p175 ICP10 protein), or anti-gD neutralized viruses (m.o.i. 10) and cell extracts obtained at 30 min p.i. were immunoblotted with ICP10 antibody that recognizes ICP10, p95 and p175 (Aurelian and Smith, 2000). ICP10 (Fig. 28, lane 3), p95 (Fig. 28, lane 4), and p175 (Fig. 28, lane 5) were seen in the respective cell extracts, but they were not seen in extracts of cells infected with antibody neutralized virus, as shown for HSV-2 in Fig. 28, lane 2). This is consistent with the failure of neutralized virus to penetrate the cells (Highlander *et al.*, 1987; Johnson *et al.*, 1990) and suggests that cell penetration occurs within 30 min p.i. The findings that: a) ICP10 is present in extracts of HSV-2 infected cells at 30 min p.i., and b) MEK/ERK activation does not require de novo viral protein synthesis, collectively suggest that ICP10 is released into the cell upon virion uncoating. However, the exact interpretation for the lower levels of p95 (relative to ICP10 and p175) in hippocampal cultures, is still unclear.

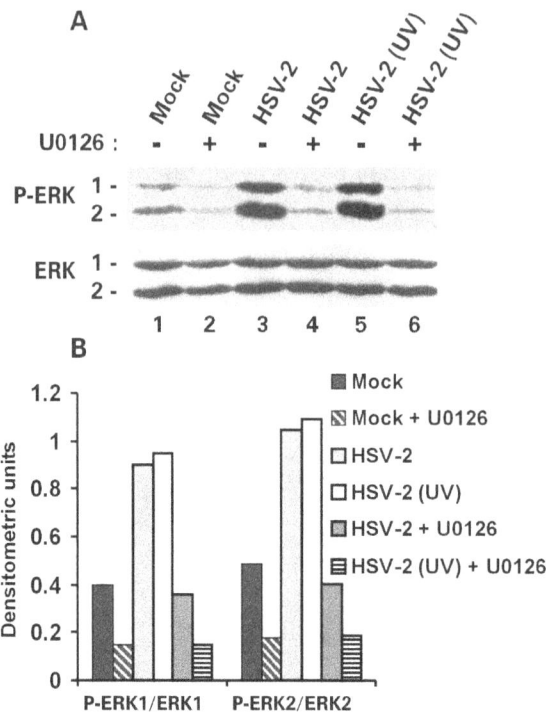

Fig. 27   **ERK activation in HSV-2-infected hippocampal neurons does not require**

de novo **viral protein synthesis. (A)** Extracts of cells mock-infected (lanes 1, 2) or

infected with HSV-2 (lanes 3, 4) or UV-inactivated HSV-2 (lanes 5, 6) in the absence

(lanes 1, 3, 5) or presence of 20 μM of U0126 (lanes 2, 4, 6) were obtained at 30 min p.i.

and immunoblotted with antibody specific for P-ERK1,2, and ERK1,2. Lanes are

identified with numbers at the bottom of the figure. **(B)** The bands in panel A were

analyzed by densitometric scanning and the results expressed as

P-ERK/ERK for both isoforms.

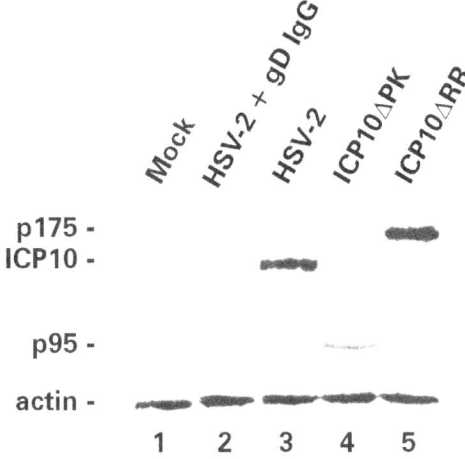

Fig. 28.   ICP10 and its mutants are present in hippocampal cells extracts at 30 min

p.i. Extracts from hippocampal cultures mock-infected (lane 1) or infected (m.o.i 10) with

HSV-2 neutralized with gD MAb (lane 2), HSV-2 (lane 3), ICP10ΔPK (lane 4) or

ICP10ΔRR (lane 5) were immunoblotted with ICP10 antibody. Blots were stripped and

reprobed with actin antibody as control for gel loading. Lanes are identified by numbers at

the bottom of the figure.

4.   Cell penetration is required for HSV-2-mediated activation of MEK/ERK in hippocampal neurons. The finding that MEK/ERK pathway is activated in hippocampal cells at 30 min (but not 24 hrs) p.i. is amenable to two potential interpretations. According to the first interpretation, MEK/ERK activation and anti-apoptotic activity are mediated by the ICP10 PK located in the tegument of the incoming virion (Smith *et al.*, 1997). Implicit in this interpretation is the assumption that cellular penetration and virion uncoating are required for both processes. An alternative interpretation is that MEK/ERK are activated by virus binding to receptors on the cell surface, and both it and the resulting anti-apoptotic effect are independent of cellular penetration. This possibility is particularly significant because an HSV receptor can generate a signal that regulates the transcription factor AP-1 upon ligand binding (Marsters *et al.*, 1997). To determine whether MEK/ERK activation is dependent on cellular penetration, cultures of hippocampal neurons were infected with antibody neutralized HSV-2 that can attach, but does not penetrate the cells (Highlander *et al.*, 1987; Johnson *et al.*, 1990). They were examined for ERK activation by immunoblotting with antibody specific for P-ERK1/2 at 30 min p.i. P-ERK1,2 levels were significantly lower in cells exposed to virus neutralized with gD MAB (Fig. 29 A, lane 3; 29B) or HSV-2 antiserum (Fig. 29 C, lane 3; 29D) than non-neutralized virus (Fig. 29A, lane 2; 29B) or virus treated with preimmune serum (Fig. 29 C, lane 2; 29 D). The superior effect of the anti-HSV-2 serum relative to the gD MAB presumably reflects the broader antigenic specificity of the antiserum. The data suggest that cell penetration is required for MEK/ERK activation.

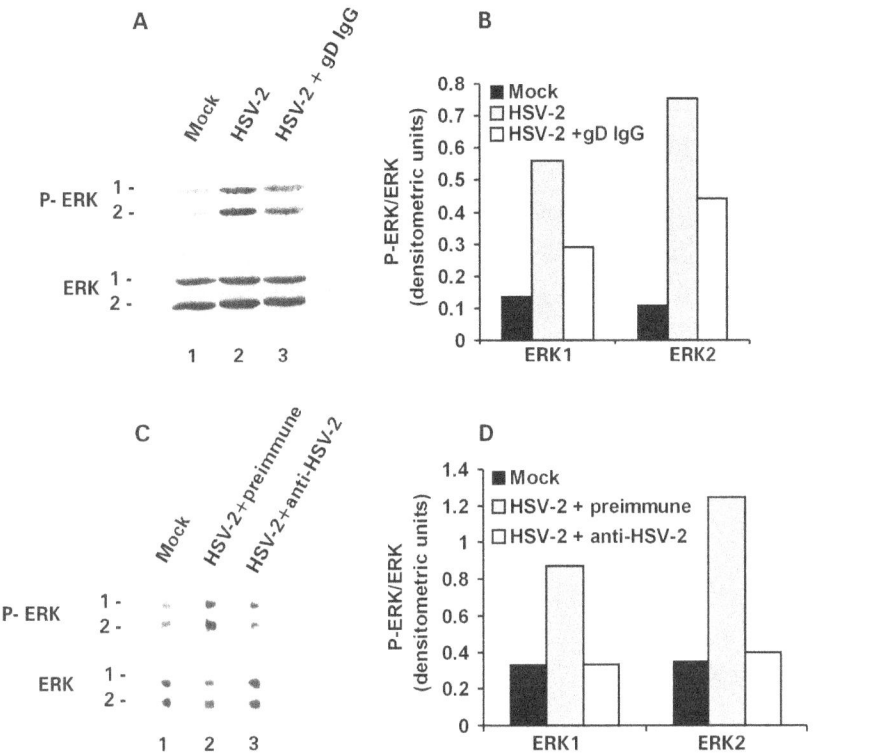

Fig. 29. **Cell penetration is required for HSV-2 mediated activation of**

MEK/ERK in hippocampal neurons. **(A)** Extracts from cells mock-infected (lane 1) or

infected with HSV-2 (lane 2) or HSV-2 neutralized with the IgG fraction from a gD MAB

(lane 3) were immunoblotted with antibody specific for P-ERK1,2 (upper bands) or

ERK1,2 (bottom bands). Lanes are identified by numbers at the bottom of the panel. **(B)**

Densitometric scanning of bands in panel **A**. Ratios for P-ERK1/ERK1 (1) and P-

ERK2/ERK2 (2) are shown. **(C)** Extracts from cells mock infected (lane 1) or infected

with HSV-2 neutralized with preimmune (lane 2) or HSV-2 hyperimmune (lane 3) serum

were immunoblotted with antibody specific for P-ERK1,2 (upper bands) or ERK1,2

(bottom bands). Lanes are identified by numbers at the bottom of the panel. **(D)**

Densitometric scanning of bands in panel **C** expressed as P-ERK1/ERK 1(1) and P-

ERK2/ERK2 (2) ratios.

5.   ICP10 PK upregulates the P-Akt levels during virus infection but does not require it for its anti-apoptotic activity. PI-K/Akt pathway is also involved in neuronal survival (Kaplan and Miller, 2000). Two series of experiments were done to examine the effect of ICP10 PK on Akt activation. In the first series of experiments, primary hippocampal cultures were infected with HSV-2 or ICP10ΔPK (m.o.i. 10) or mock-infected, and cell extracts were examined at 0.5 hrs p.i. by immunoblotting with antibodies specific for activated Akt (P-Akt) and total Akt. As shown in Fig. 30 A, B, both viruses induce increased expression of Akt (Fig. 30 A, lanes 2 and 3, for ICP10ΔPK and HSV-2, respectively) compared to the levels seen in mock infected cells (Fig. 30 A, lane 1). This effect is not due to improper gel loading because actin levels were similar in all samples (data not shown). Both ICP10ΔPK and HSV-2 viruses trigger Akt activation at similar levels, as seen in the blot from Fig. 30 A (upper panel, lanes 2 and 3, respectively) and the densitometric scanning analysis of P-Akt/Akt ratios (Fig. 30 B), suggesting that other viral genes, but not ICP10 PK, are involved in Akt activation at 0.5 hrs p.i. In the second series of experiments, primary cultures of hippocampal neurons were mock-infected, or infected with HSV-2, ICP10ΔPK, ICP10ΔRR or HSV-1 (m.o.i. 10). To determine whether Akt activation/phosphorylation depends on PI3-K, cultures were infected with HSV-2 presence of 100 μM of the specific PI3-K inhibitor LY294002 (Vlahos *et al.*, 1994). Cell extracts were examined at 24 hrs p.i. by immunoblotting with P-Akt and Akt antibodies. Increased Akt levels (Fig. 30 C, middle panel) were observed in cells infected with HSV-2 (lane 3), ICP10ΔPK (lane 4) and ICP10ΔRR (lane 5), but not HSV-1 (lane 6), suggesting that the viral gene responsible for this effect is serotype (HSV-2)-specific. However, the significance of this effect is unclear. Akt phosphorylation/activation was not increased in cells infected with ICP10ΔPK (Fig. 30 C, lane 4) or HSV-1 (Fig. 30 C, lane 6) compared to the levels seen in mock-infected cells (Fig. 30 C, lane 1). By contrast, Akt was activated in cells infected with HSV-2 and ICP10ΔRR (Fig. 30 C, lanes 3 and 5, respectively), suggesting that ICP10 PK is involved in upregulation of P-Akt levels. The Akt activation seen in HSV-2-infected cultures was PI3-K dependent because the specific PI3-K inhibitor LY294002 completely inhibited it (Fig. 30 C, lane 2). The conclusion that Akt is activated (by a PI3-K-dependent mechanism) with ICP10 PK-expressing viruses (HSV-2 and ICP10ΔRR) but not ICP10ΔPK or HSV-1, is supported by densitometric scanning and data expression as P-Akt/Akt ratios (Fig. 30 D). To test whether PI3-K/Akt is involved in cell survival, primary cultures of hippocampal neurons were infected with HSV-2 (m.o.i. 10) or growth medium (mock) for 24 hrs in the absence or presence of increasing concentrations (10-100 μM) of the specific PI3-K inhibitor LY294002, and analyzed by TUNEL. As shown in Fig. 31, LY294002 had a minimal effect on the % TUNEL-positive cells in HSV-2 infected cultures (9.1 $\pm$ 1 % and 24.8 $\pm$ 2.3 % at 0 and 100 μM, respectively), suggesting that this pathway has little contribution to the survival of HSV-2-infected hippocampal neurons. By contrast, LY294002 had a major effect on the survival of mock-infected cells, with 5 $\pm$ 3 % apoptotic cells in the untreated cultures as compared to 72 $\pm$ 4.5 at 100 μM of LY294002 (p<0.01 by ANOVA), suggesting that the PI3K/Akt pathway is required for basal maintenance of hippocampal neurons. Similar results were obtained with increasing concentration of the PI3-K inhibitor wortmannin (data not shown). Presumably, HSV-2 infection uncouples cell survival from PI3-K activation by activating the MEK/ERK pathway and inhibiting the pro-apoptotic pathways (such as JNK/c-Jun) which are activated upon PI3-K/Akt inhibition, as it will be discussed further. However, the possibility that both pathways converge upon similar downstream effectors cannot be excluded.

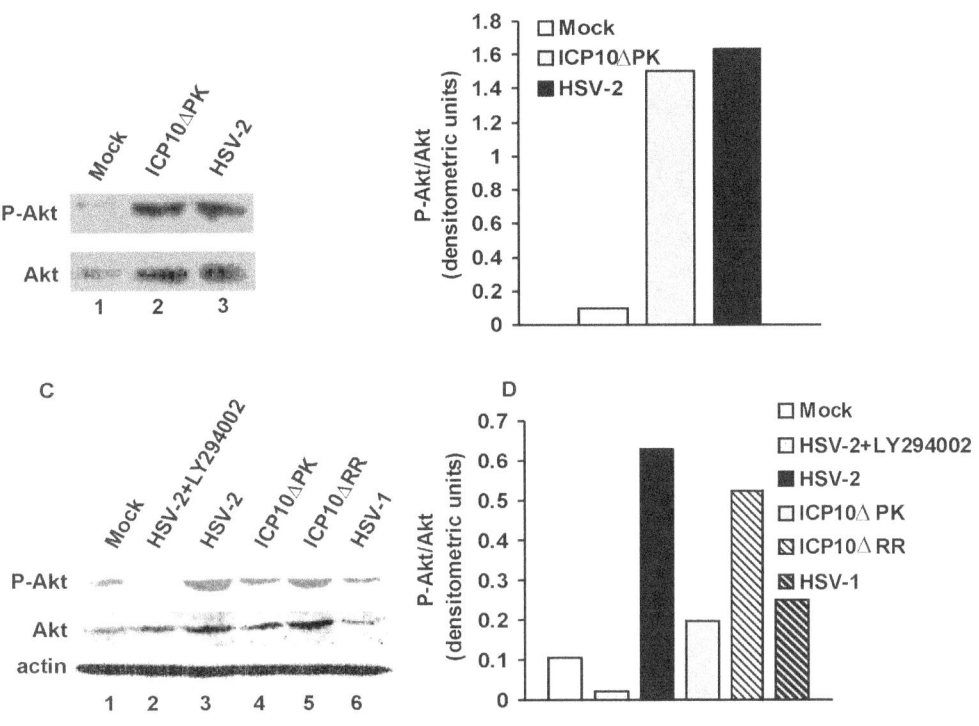

Fig. 30.   ICP10 PK upregulates the P-Akt levels during virus infection. (A) Primary

hippocampal cultures were infected with 10 pfu/cell of HSV-2 (lanes 3), ICP10ΔPK (lanes

2), or mock-infected (lane 1) and analyzed at 0.5 hrs p.i. by immunoblotting with

antibodies specific for P-Akt (phosphorylated/activated Akt) (upper panel) or total Akt

(lower panel). Lanes are identified by numbers at the bottom of the panel. **(B)**

Densitometric scanning analysis of bands from (A) with results expressed as P-Akt/Akt

ratios. **(C)** Primary hippocampal cultures were infected with 10 pfu/cell of HSV-2 [in the

absence (lane 3) or presence (lane 2) of LY294002], ICP10ΔPK (lane 4), ICP10ΔRR (lane

5) or HSV-1 (lane 6), or mock-infected (lane 1) and analyzed at various times p.i. by

immunoblotting with antibodies specific for P-Akt (phosphorylated/activated Akt) (upper

panel), total Akt (middle panel), and action (bottom panel) for loading control. Lanes are identified by numbers at the bottom of the panel. **(D)** Densitometric scanning analysis of

bands from (C) with results expressed as P-Akt/Akt ratios.

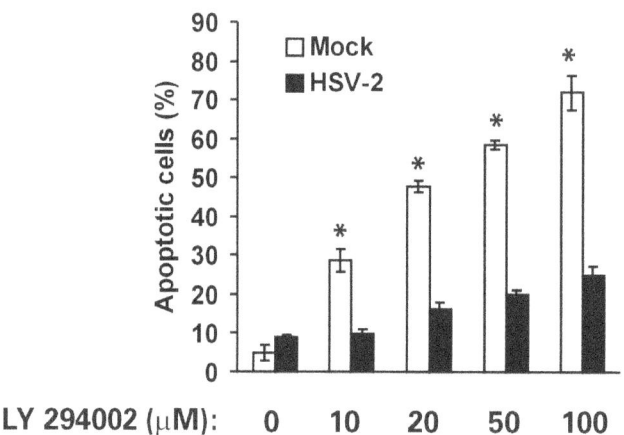

Fig. 31 **PI3-K/Akt pathway activation is not required for survival of HSV-2-**

infected hippocampal neurons. Primary hippocampal cultures were infected for 24 hrs

with HSV-2 or mock-infected in the absence or presence of increasing concentrations of

LY294002 (10-100 μM) and analyzed by TUNEL. Results shown are average of three

independent experiments, and are expressed as % apoptotic cells $\pm$ SEM. *, $p<0.01$ vs

HSV-2, by ANOVA.

6.   ICP10 PK induces CREB phosphorylation and stabilizes it. CREB is a 43 kDa (p43) transcription factor regulated by phosphorylation at Ser133 (located within the N-terminal transactivation domain) through both the Ras/Raf/MEK/ERK/Rsk and the PI3-K/Akt pathways (Finkbeiner *et al.*, 1997; François *et al.*, 2000). CREB regulates many aspects of neuronal function such as neuronal excitation, CNS development and long term memory formation and is important for neurotrophin-mediated gene transcription associated with neuronal survival. For example, CREB appears to be a primary transcriptional activator of the anti-apoptotic gene Bcl-2 (Riccio *et al.*, 1999). In order to determine whether ICP10 PK regulates CREB phosphorylation, extracts of cells infected with HSV-2, ICP10ΔPK or mock-infected cells, were immunoblotted with antibodies specific for Ser133-phosphorylated CREB (P-CREB) (Fig 32, upper panel), total CREB (Fig 32, middle panels) or actin (for loading control, Fig.32, bottom panel). CREB phosphorylation/activation was equally induced at 0.5 hrs p.i. by HSV-2 (lane 2) and ICP10ΔPK (lane 3). A significant difference in CREB phosphorylation starts to be seen in HSV-2-infected cells at 4 hrs pi (lane 4) and 8 hrs p.i. (lane 6) compared to ICP10ΔPK (lanes 5 and 7, respectively), which may suggest that *de novo* synthesized ICP10 PK is responsible for inducing CREB phosphorylation. ICP10 PK also stabilizes P-CREB as evidenced by its expression at 24 hrs p.i. in HSV-2 (lane 8) but not ICP10ΔPK-infected cells (lane 9). Significantly, immunoblotting with the CREB antibody resolved a 30 kDa band in the ICP10ΔPK-infected cell extracts (at 24 hrs p.i.) in addition to the p43 CREB protein (Fig. 32, lane 9). The p30 band is consistent with the previously reported CREB cleavage product that occurs in response to caspase activation (François *et al.*, 2000), suggesting that in ICP10ΔPK-infected cells, CREB is inactivated by multiple mechanisms such as dephosphorylation and proteolysis. The p30 band was not seen in mock or HSV-2-infected cell extracts at any time points tested (Fig. 32), suggesting that CREB is not a target for proteolytic destruction in these cells. These data suggest that ICP10 PK may promote survival of neurons, through CREB-dependent mechanisms.

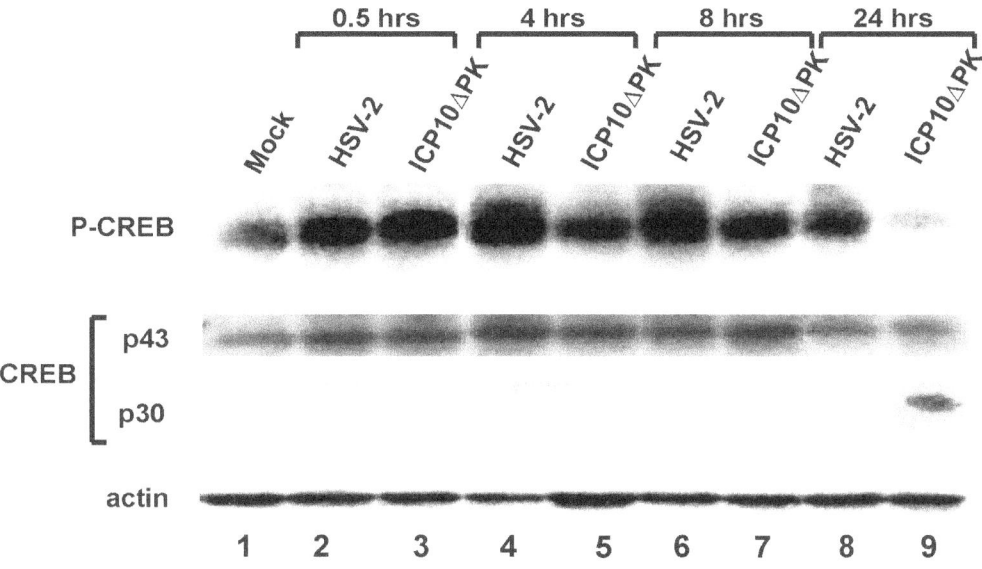

Fig. 32. ICP10 PK activates and stabilizes CREB. Primary hippocampal cultures were

infected with HSV-2 (lanes 2, 4, 6, 8), ICP10ΔPK (lanes 3, 5, 7, 9), or mock infected (lane

1) and analyzed by immunoblotting with antibody specific for phosphorylated

(Ser133)/activated CREB (P-CREB) (upper panel). Blots were stripped twice and

reprobed with antibody specific for total CREB (middle panel) or actin (bottom panel)-as

control for gel loading. Lanes are identified by numbers at the bottom of the figure.

7.   PI3-K/Akt and α7 nAChR are not involved in CREB activation in HSV-2-infected hippocampal neurons. In addition to the Raf/MEK/ERK pathway, PI3-K/Akt also participates in CREB activation (Brunet *et al.*, 1999), suggesting that CREB may be a common downstream effector for these two pathways in regulating the survival of HSV-2-infected hippocampal neurons. Moreover, as discussed in Section I Chapter 1.2, MEK/ERK and PI3-K/Akt pathway contribute to CREB activation/phosphorylation and are important effectors of cholinergic function (Messer *et al.*, 1991). Nicotine (agonist for nAChR) activate ERK2 in hippocampal cells by an α7 nAChR-dependent mechanism, because an α7 nAChR selective antagonist, methylcaconitine inhibits nicotine's effect on ERK2 (Dineley *et al.*, 2001). Moreover, α7 nAChR was associated with PI3-K and activates the PI3-K/Akt pathway as a mechanism for protection from apoptosis (Kihara et al., 2001). To examine the role of PI3-K/Akt and α7 nicotinic acetylcholine receptor (α7 nAChR) signaling in CREB activation induced by HSV-2, two series of experiments were done. In the first series of experiments, primary cultures of hippocampal neurons were infected with HSV-2 or mock-infected for 24 hrs, in the absence or presence of LY294002 (100 μM) or the α7 nAChR antagonist α-bungarotoxin (αBGTX, 100 nM) and cell extracts were examined by immunoblotting with antibodies specific for P-CREB or CREB. As shown in Fig. 33 (A, B), P-CREB levels in HSV-2-infected cultures in the absence (Fig. 33 A, lane 4, Fig. 33 B) or presence of LY294002 (Fig. 33 A, lane 5, Fig. 33 B) or αBGTX (Fig. 33A, lane 6, Fig. 33B) were similar and higher than those observed in mock-infected cultures (lane 1), suggesting that PI-3K/Akt and α7 nAChR signaling are not involved in HSV-2-induced CREB activation. By contrast, LY294002 treatment of mock-infected cultures reduced the level of P-CREB (Fig. 33 A, lane 2, Fig. 33 B) while αBGTX

fully inhibited CREB activation (Fig. 33 A, lane 3, Fig. 33 B), suggesting that in non-

infected hippocampal neurons, CREB activation (and therefore the specific gene

transcription it mediates) is due to, at least in part, to PI3-K/Akt and α7 nAChR signaling.

In view of the complete inhibition of CREB activation by αBGTX in non-infected

hippocampal neurons, it seemed of interest to examine whether αBGTX also affects their

survival. Thus, in the second series of experiments, hippocampal cultures were pre-treated

(or not) with 100 nM αBGTX and mock- or HSV-2-infected in the absence or presence of

the antagonist. αBGTX-treatment increased the cell death in mock-infected cultures from

7.3 ± 2.5 % (not treated) to 24.8 ± 3.3 % (in the presence of αBGTX) ($p<0.05$ by

ANOVA) (Fig. 34). Notwithstanding, αBGTX treatment also increased the number of

apoptotic cells in HSV-2-infected cultures, from 11.3 ± 2.4 % to 28.6 ± 1.8 % ($p<0.05$, by

ANOVA) (Fig. 34), suggesting that α7 nAChR signaling [PI-3K/Akt and/or MEK/ERK

(Dineley et al., 2001; Kihara et al., 2001)] is responsible, at least in part, for the survival

of hippocampal neurons, regardless of their infection status. However, the PI3-K/Akt

pathways is not required for survival of HSV-2-infected hippocampal neurons (as shown

in Fig. 31) or HSV-2 induced CREB phosphorylation/activation (Fig. 33), suggesting that

αBGTX triggers apoptosis by inhibiting another survival pathway, most likely MEK/ERK.

By contrast, inhibition of CREB activation in αBGTX-treated mock-infected hippocampal

neurons suggests that, in these cultures, αBGTX may induce apoptosis by interfering with

CREB-mediated transcription of anti-apoptotic genes.

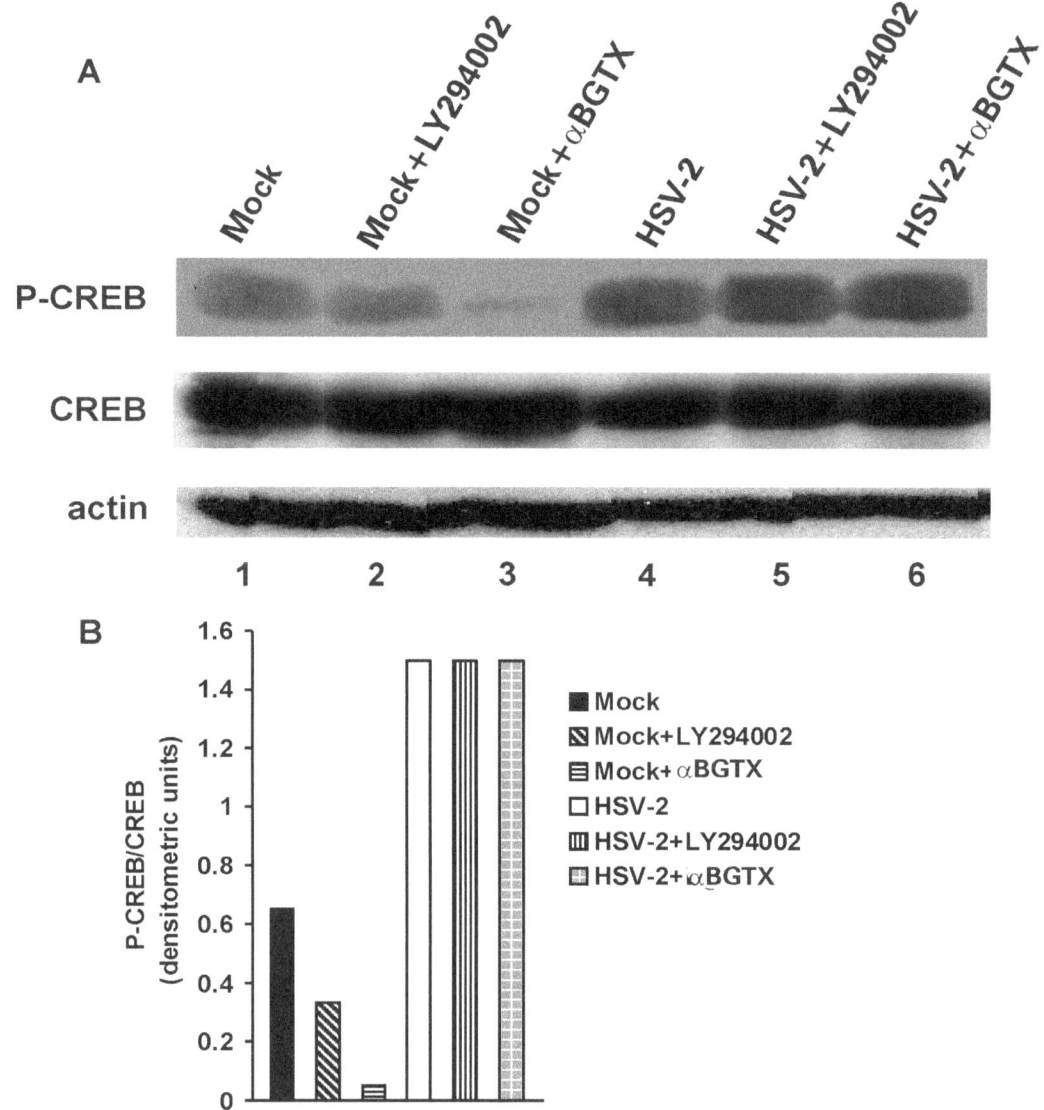

Fig. 33. PI3-K/Akt and nAChR are not involved in CREB activation in HSV-2-infected hippocampal neurons. **(A)** Primary cultures of hippocampal neurons were infected with HSV-2 (lanes 4, 5. 6), or mock infected (lanes 1, 2, 3) in the absence (lanes 1 and 4, for mock and HSV-2, respectively) or presence of α-bungarotoxin (αBGTX, 100 nM) (lanes 3 and 6, for mock and HSV-2, respectively) or LY294002 (100 μM) (lanes 2 and 5, for mock or HSV-2, respectively). Immunoblotting was with antibody specific for

phosphorylated (Ser133)/activated CREB (P-CREB) (upper panel). Blots were stripped twice and reprobed with antibody specific for total CREB (middle panel) or actin (bottom panel)-as control for gel loading. Lanes are identified by numbers at the bottom of the figure.(B) Densitometric scanning of P-CREB and CREB bands from (A). Results are expressed as P-CREB/CREB ratios.

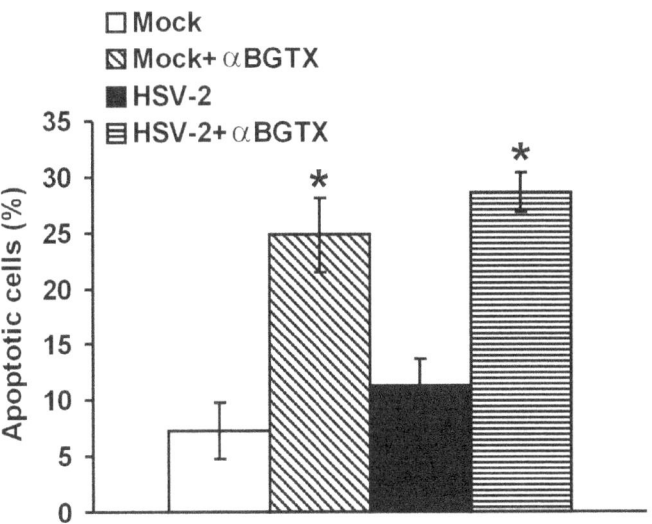

Fig. 34   nAChR signaling is involved in survival of hippocampal neurons. Primary cultures of hippocampal neurons were infected with HSV-2, or mock infected for 24 hrs in the absence or presence of 100 nM αBGTX and analyzed by TUNEL. Results are average of three independent experiments, and are expressed as % apoptotic cells $\pm$ SEM. *, $p < 0.05$ vs not-treated mock- or HSV-2-infected neurons, by ANOVA

8.   **Inhibition of PI3-K/Akt and α7 nAChR signaling is associated with c-Jun activation.** Experiments described in this report showed that mock-infected and HSV-2-infected hippocampal neurons differ in their undertaking of specific signaling pathways required for survival. Thus, inhibition of PI3-K/Akt by the specific PI3-K inhibitor LY294002 caused extensive cell death in mock-infected hippocampal neuron but not in HSV-2 infected cultures (Fig. 31). In addition, the α7 nAChR antagonist, αBGTX , induced apoptosis in both mock- and HSV-2-infected cultures (Fig. 34), but inhibited CREB activation only in mock-infected cells, suggesting that there are distinct mechanisms by which αBGTX induces apoptosis in mock- or HSV-2-infected cells. To examine whether LY294002 and αBGTX-induced cell death requires activation of the apoptosis-associated transcription factor c-Jun, primary hippocampal cultures were infected with HSV-2 (m.o.i. 10) or mock-infected for 24 hrs in the absence or presence of 100 μM LY294002 or 100 nM αBGTX, and c-Jun activation was analyzed by immunoblotting as described above. As seen in Fig. 35 A, LY294002 treatment of mock-infected cultures (lane 2) induced significant activation of c-Jun, as evidenced by increased levels of P-Jun (Ser 73) and especially P-Jun (Ser 63) compared to non-treated cultures (Fig. 35 A, lane 1). Densitometric scanning and analysis of data as % treated vs untreated suggests that LY294002 caused 320 % increase in P-Jun (Ser 73) activation (a 3.2 fold) (Fig. 35 B) and 516 % increase (a 5.2 fold) in P-Jun (Ser 63) activation (Fig. 35 C). LY294002 treatment also induced c-Jun activation in HSV-2-infected cultures (Fig. 35 A, lane 5), of 178 % (a 1.8 fold) for P-Jun (Ser 73) (Fig. 35 B) and 346 % (a 3.5 fold) for P-Jun (Ser 63) (Fig. 35 C). To summarize, inhibition of PI3-K/Akt pathway by LY294002, in mock-infected hippocampal neurons, is associated with activation of c-Jun (Fig. 35) and inhibition of CREB (Fig. 33) transcription factors and leads to apoptosis (Fig. 31). By contrast, inhibition of PI3-K/Akt pathway by LY294002, in HSV-2-infected hippocampal neurons, is associated with activation of c-Jun (Fig. 35) but has no effect on HSV-2-induced CREB activation (Fig. 33) and does not affect cell survival (Fig. 31), suggesting that activation of CREB is the key factor responsible for the distinct fate of infected versus non-infected hippocampal neurons. αBGTX treatment of mock-infected cultures (Fig. 35 A, lane 3) induced activation of c-Jun, as evidenced by increased levels of P-Jun (Ser 63) [but less of P-Jun (Ser 73)] compared to non-treated cultures (Fig. 35 A, lane 1). Densitometric scanning and analysis of data as % treated vs untreated suggests that αBGTX caused 210 % increase in P-Jun (Ser 73) activation (a 2.1 fold) (Fig. 35 B) and 296 % increase (a 3 fold) in P-Jun (Ser 63) activation (Fig. 35 C). In the same manner, αBGTX treatment induced c-Jun activation in HSV-2 infected and treated cultures (Fig. 35A, lane 6), of 129 % (a 1.3 fold) for P-Jun (Ser 73) (Fig. 35 B) and 227 % (a 2.3 fold) for P-Jun (Ser 63) (Fig. 35 C), suggesting that αBGTX-induced inhibition of α7 nAChR signaling leads to c-Jun activation, which may be responsible for the increase in the numbers of apoptotic cells upon αBGTX treatment (shown in Fig. 34). A potential interpretation on why αBGTX-induced c-Jun activation is not inhibited in HSV-2-infected cultures (that express ICP10 PK) may be that αBGTX interferes with, or inhibits MEK/ERK survival pathway that HSV-2-infected neurons depend on for their survival. In this respect, αBGTX treatment has a similar effect on survival of HSV-2-infected hippocampal neurons as other inhibitors of MEK/ERK pathway described in this report, specifically, induction of cell death. Similarly, the increase in c-Jun activation and cell death observed in mock-infected cultures upon treatment with αBGTX may be due to inhibition of PI3-K/Akt signaling which is the main survival pathway in non-infected hippocampal neurons. The relative small increase in c-Jun activation induced by αBGTX in mock-infected cultures, compared to those induced by LY294002, are in accordance with the relative low % of apoptosis (25- 30 %) (Fig. 34) induced by αBGTX compared to LY294002 (~ 75 %) (Fig. 31). Similarly, αBGTX was less potent than LY294002 in inducing c-Jun activation in HSV-2-infected cultures (Fig. 35 A), but apoptosis levels were similarly low for αBGTX- or LY294002-treated cultures (~ 25-30%) (Fig. 34 and Fig. 31, respectively), suggesting that: i) αBGTX may partially interfere with MEK/ERK pathway activation (but not CREB activation) and the virus ability to counteract c-Jun activation; and ii) inhibition of PI3-K/Akt pathway by LY294002 (associated with increased c-Jun activation) is counteracted by ICP10 PK-induced MEK/ERK activation and CREB activation/stabilization.

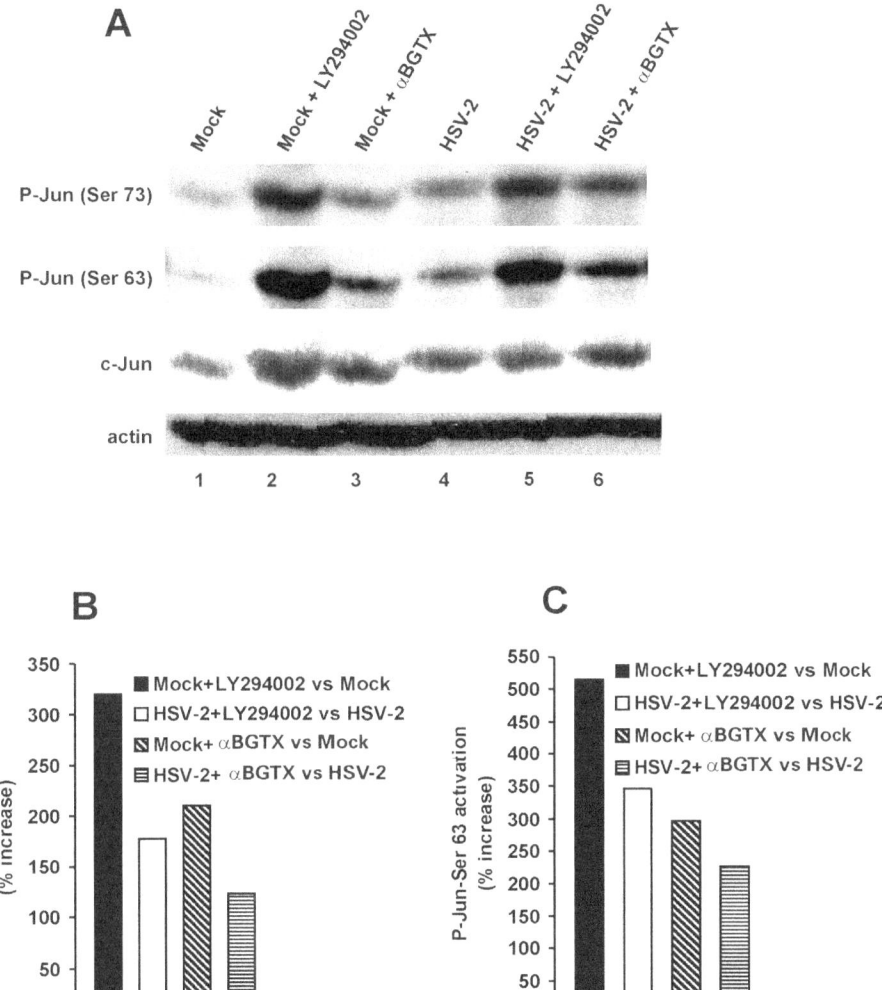

Fig. 35. **Inhibition of PI3-K/Akt and α7 nAChR signaling is associated with c-Jun**

activation. (A) Proteins from extracts of hippocampal cells infected with HSV-2 (m.o.i.

10) or mock-infected for 24 hrs (in the presence or absence of 100 μM LY294002, or 100

nM αBGTX), were resolved by SDS-PAGE, transferred to nitrocellulose membrane, and

immunoblotted with antibodies specific for c-Jun phosphorylated (P-Jun) at Ser 73 (upper

panel) or Ser 63 (second from the top panel), c-Jun (recognizes c-Jun independent of its state of phosphorylation) (third panel) and actin (bottom panel, for loading control). Lanes

are identified by numbers at the bottom of the panel. **(B)** Densitometric scanning and data

analysis of bands from panel A as % P-Jun (Ser 73) activation (treated sample versus non-

treated x 100). **(C)** Densitometric scanning and data analysis of bands from panel A as %

P-Jun (Ser 63) activation (treated sample versus non-treated x 100).

9. **HSV-2 (but not HSV-1) stabilizes the Bcl-2 expression levels.** One of the key genes involved in protection from apoptosis is Bcl-2, whose transcription is CREB-dependent (Kaplan and Miller, 2000). The inhibition of apoptosis in the HSV-2- versus HSV-1-infected hippocampal cultures may reflect activation of endogenous protective mechanisms, such as an increase in the Bcl-2 expression levels which is associated with neuronal survival (Tsujimoto, 1998). To examine the kinetics of Bcl-2 protein expression during HSV infection of hippocampal neurons, cultures were infected with HSV-2 or HSV-1 (m.o.i. 10), or mock-infected, and analyzed at various times p.i. by immunoblotting with antibody specific for Bcl-2. As seen in Fig. 36, at 0.5 hrs p.i., the Bcl-2 expression in HSV-1-infected cultures (lane 3) was decreased compared to HSV-2 (lane 2) or mock (lane 1) but increased at 6 hrs p.i. (lane 5) [to the same extent as that induced by HSV-2 (lane 4)] only to fall below detectable levels at 24 hrs p.i. (lane 7). By contrast, Bcl-2 expression was still seen at 24 hrs p.i. in HSV-2 infected cultures (Fig. 36, lane 6) (albeit at lower levels than those seen at 6 hrs p.i.), suggesting that the two HSV types differ in their ability to activate endogenous survival mechanisms that involve the Bcl-2 protein. At this time we do not know which viral genes (most likely from the immediate early class) are responsible for this transient increase in Bcl-2 levels which is seen at 6 hrs p.i. However, the decrease in Bcl-2 levels triggered by HSV-1 infection at 24 hrs p.i. may be due specific cleavage of Bcl-2 by caspases, which leads to inactivation of its anti-apoptotic function (Grandgirard *et al.*, 1998). Significantly, as already described, at this time p.i. (24 hrs) more than 55 % of cells exhibit DNA fragmentation, caspase activation and PARP cleavage. However, no Bcl-2 cleavage products were detected, most likely due to a rapid degradation/clearance by additional proteases. Alternatively, the

destabilization of the Bcl-2 protein may be due to activation of the pro-apoptotic JNK

pathway, as previously described for p38 MAPK (Zachos *et al.*, 2001). An alternative

explanation is that, in HSV-2 - but not HSV-1-infected hippocampal neurons, the Bcl-2

protein is stabilized by its interaction with specifically-induced chaperone proteins, such as

Bag-1.

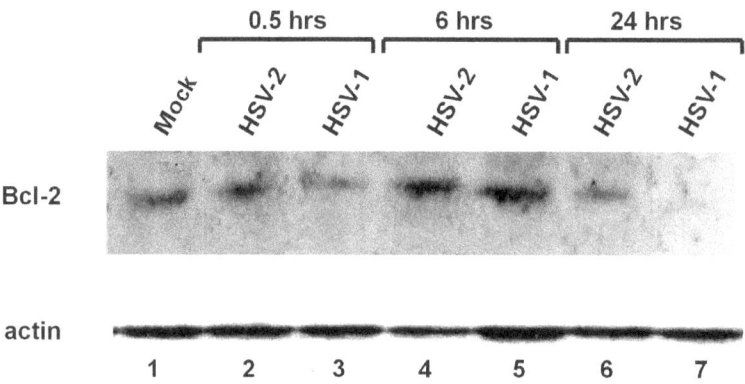

Fig. 35. **Bcl-2 expression in HSV-infected hippocampal neurons.** Proteins from

extracts of hippocampal cells either mock-infected (lane 1) or infected with HSV-1 (lanes

3, 5, 7) or HSV-2 (lanes 2, 4, 6) for the indicated times, were resolved by SDS-PAGE,

transferred to nitrocellulose membrane, and immunoblotted with antibodies specific for

Bcl-2 (upper panel) or actin (bottom panel). Lanes are identified by numbers at the bottom

of the figure.

10. **ICP10 PK induces the expression of the anti-apoptotic protein Bag-1.** Bcl-2

prevents many, but not all, forms of apoptotic cell death (Tsujimoto, 1998) which suggests

the existence of multiple, independent intracellular mechanisms of apoptosis.

Alternatively, these additional pathways may involve proteins that differentially regulate

Bcl-2 function (Boise *et al.*, 1993). Such a protein is Bag-1, which interacts physically and

functionally with Bcl-2 to cooperatively interfere with the apoptotic cascade at the level of

caspase activation (Takayama *et al.*, 1995; Bardelli *et al.*, 1996; Schulz *et al.*, 1997). Not

only does it enhance anti-apoptotic activity of Bcl-2 but also inhibits apoptosis by itself

(Takayama *et al.*, 1995). In order to determine whether HSV infection (and specifically the

ICP10 PK gene) affects Bag-1 expression levels, hippocampal cultures were infected with

HSV-1, HSV-2, ICP10ΔPK or ICP10ΔRR, or mock-infected, and Bag-1 expression was

determined by immunoblotting. Bag-1 expression was not seen in mock-infected cultures

(Fig. 37 A, lane 1) nor in cultures infected with the apoptotic viruses HSV-1 or ICP10ΔPK

(Fig. 37 A, lanes 3,7 and 11 for ICP10ΔPK; and lanes 5, 9 and 13 for HSV-1). By contrast,

HSV-2 infection induced the expression of Bag-1 starting around 6 hrs p.i. (Fig. 37 A, lane

6) and increasing by 24 hrs p.i. (lane 10). The viral gene responsible for this effect was

ICP10 PK because Bag-1 expression was also induced by the ICP10ΔRR mutant which

retains the ICP10 PK DNA, at same levels at HSV-2 by 6 hrs p.i. (Fig. 37 A, lane 8) but at

lower levels than HSV-2 by 24 hrs p.i. (Fig. 37 A, lane 12). In order to determine the

mechanisms of Bag-1 regulation, hippocampal cultures were infected with HSV-2 for 24

hrs in the absence or presence of the specific MEK1,2 inhibitor U0126 (Favata et al.,

1998), c-Raf kinase inhibitor I (Lackey *et al.*, 2000), PI-3K inhibitor LY294002 (Vlahos *et al.*, 1994) or p38 MAPK inhibitor SB203580 (Cuenda *et al.*, 1995) at 20, 50, 50 and 10 µM, respectively (Fig. 37 B) . The SB203580 inhibitor (Fig. 37 B, lane 6) only minimally affected the Bag-1 expression. By contrast, the c-Raf kinase inhibitor I (Fig. 37 B, lane 4), U0126 (Fig. 37 B, lane 2), and LY294002 (Fig. 37 B, lane 5) inhibited the HSV-2-induced expression of Bag-1, suggesting that both Raf/MEK/ERK and PI3-K/Akt pathways contribute to its regulation.

169

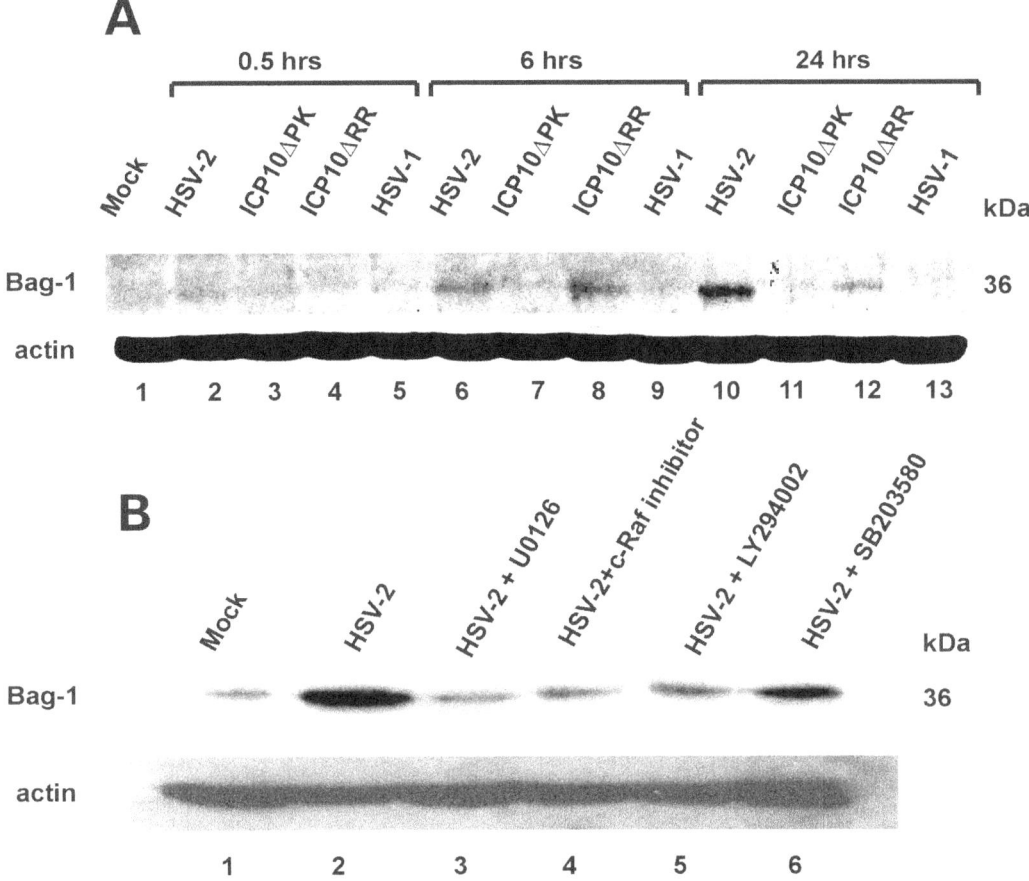

Fig. 37. **HSV-2 gene ICP10 PK induces the expression of Bag-1 in hippocampal**

neurons. (A) Primary hippocampal cultures were either mock-infected or infected with

HSV-1, HSV-2, ICP10ΔPK or ICP10ΔRR for the indicated times. Proteins were resolved

by SDS-PAGE, transferred to nitrocellulose membrane, and immunoblotted with antibody

specific for Bag-1 (upper panel) or actin (bottom panel). **(B)** Primary hippocampal cultures

were either mock-infected, or infected with HSV-2 for 24 hrs in the absence or presence of

U0126 (20μM), c-Raf-1 kinase inhibitor I (50μM), LY294002 (100μM) or SB203580

(10μM). Bag-1 expression (upper panel) was determined by immunoblotting with specific

antibody. Actin (bottom panel) is shown for loading control. Lanes are identified by

numbers at the bottom of each panel.

11.   Ectopically expressed Bag-1 protein inhibits virus-induced apoptosis in

hippocampal neurons. Bag-1 was previously reported to interact physically and

functionally with Bcl-2 and cooperatively interfere with the apoptotic cascade at the level

of caspase activation (Bardelli *et al.*, 1996; Schulz *et al.*, 1997; Takayama *et al.*, 1995). It

was also shown to inhibit apoptosis by itself (Takayama *et al.*, 1995). Thus, it seemed of

interest to determine whether Bag-1 (which is specifically induced by ICP10 PK in virus-

infected hippocampal neurons) can inhibit virus-induced apoptosis. To test this hypothesis,

two series of experiments were done. In the first series of experiments, primary cultures of

hippocampal neurons were transfected with pJG4-5m Bag-1 expression vector [which

encodes the murine Bag-1 protein and was previously described (Takayama *et al.*, 1995)]

using the FuGene transfection reagent, according to the manufacturer instructions. The

efficiency of transfection (30-40 %) was determined by immunostaining with antibody

specific for Bag-1. At 48 hrs post-transfection, cultures were infected for 24 hrs with

HSV-2, HSV-1, ICP10ΔPK, or mock-infected, then stained by the immunoperoxidase

method with antibody specific for the cleaved fragment of PARP (p85PARP), the

occurrence of which is regarded as an apoptotic marker (Johnson Webb *et al.*, 1997). Non-

transfected cultures similarly infected were used as controls. Positively and negatively-

stained cells were counted and results were expressed as % p85PARP-positive cells $\pm$

SEM (Fig. 38 A). Transfection did not affect significantly (p>0.05, by ANOVA) the %

positive cells in mock- (2.3 $\pm$ 0.5) or HSV-2- (3.1 $\pm$ 2) infected cultures compared to the

% obtained for non-transfected cells (4.2 $\pm$ 1.3 and 9.9 $\pm$ 0.2 for mock and HSV-2,

respectively) (Fig. 38 A). By contrast, a very significant reduction (p<0.001) in the % of p85PARP-positive cells was seen in cultures Bag-1- transfected and infected with ICP10ΔPK (2.6 ± 1.3) or HSV-1 (6.7 ± 1.7) compared to the % obtained for similarly infected but non-transfected cultures (39.5 ± 2.5 and 33.8 ± 2.2, for ICP10ΔPK and HSV-1, respectively) (Fig. 38 A), suggesting that Bag-1 protein interferes with PARP cleavage in virus-infected hippocampal neurons. In the second series of experiments, hippocampal cultures transfected for 24 hrs with pJG4-5mBag-1 and infected as above, were analyzed by TUNEL at 24 hrs p.i. TUNEL-positive and TUNEL-negative cells were counted and results were expressed as % apoptotic cells ± SEM. Transfection did not affect significantly (p>0.05, by ANOVA) the % apoptotic cells in mock- (2.7 ± 0.6) or HSV-2- (9.2 ± 0.6) infected cultures compared to the % obtained for non-transfected cells (4.1 ± 0.6 and 6.8 ± 0.8 for mock and HSV-2, respectively) (Fig. 38 B). By contrast, a very significant reduction (p<0.001) in the % apoptotic cells was seen in cultures Bag-1-transfected and infected with ICP10ΔPK (11.8 ± 1.3) or HSV-1 (3.4 ± 0.9) compared to the % obtained for similarly infected but not transfected cultures (60.9 ± 3.8 and 57.7 ± 3.6, for ICP10ΔPK and HSV-1, respectively) (Fig. 38 B), suggesting that Bag-1 protein inhibits virus-induced apoptosis in hippocampal neurons. Collectively, these data suggest that the anti-apoptotic activity of ICP10 PK, as evidenced in virus-infected hippocampal neurons, is due, at least in part, to its specific induction of Bag-1 by a mechanism that involves activation of survival signaling pathways.

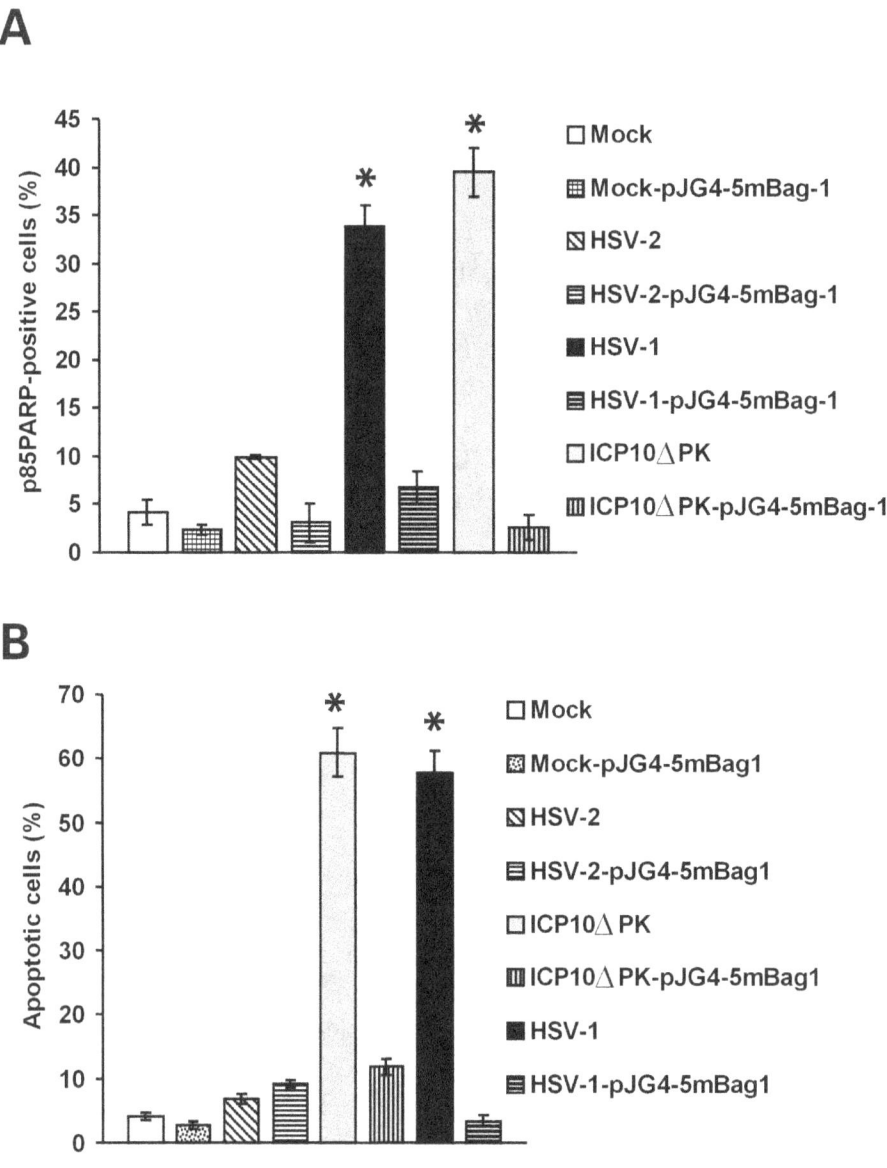

Fig. 38   Ectopically expressed Bag-1 protein inhibits virus-induced apoptosis in

hippocampal neurons. (A) Primary cultures of hippocampal neurons were transfected

with pJG4-5mBag-1 expression vector using the FuGene transfection reagent, according to the manufacturer instructions. At 48 hrs post-transfection, cultures were infected for 24 hrs with HSV-2, HSV-1, ICP10ΔPK, or mock-infected, then stained by immunoperoxidase method with antibody specific for the cleaved fragment of PARP (p85PARP). Non-transfected cultures similarly infected were used as controls. Positively and negatively-stained cells were counted and results were expressed as % p85PARP-positive cells $\pm$ SEM. Results shown are average of three independent experiments. *, $p<0.001$ versus transfected and infected (mock, HSV-1, HSV-2, ICP10ΔPK)-cultures and versus non-transfected and infected (mock, HSV-2)-cultures. **(B)** Primary cultures of hippocampal neurons were transfected with pJG4-5mBag-1 expression vector, as above. At 24 hrs post-transfection, cultures were infected for 24 hrs with HSV-2, HSV-1, ICP10ΔPK, or mock-infected, then analyzed by TUNEL. Non-transfected cultures similarly infected were used as controls. TUNEL-positive and TUNEL-negative cells were counted and results were expressed as % apoptotic cells $\pm$ SEM. Results shown are average of three independent experiments. *, $p<0.001$ versus transfected and infected (mock, HSV-1, HSV-2, ICP10ΔPK)-cultures and versus non-transfected and infected (mock, HSV-2)-cultures.

12.   Discussion-Specific Aim 2

The experiments presented in the Specific Aim 1 Section suggested that ICP10 PK protects hippocampal neurons from virus-induced apoptosis by a mechanism that involved inhibition of caspase-3 activation, PARP cleavage and DNA fragmentation. Because ICP10 PK activates the Ras/Raf/MEK/ERK pathway in non-neuronal cells (Smith *et al.*, 2000), and this pathway is associated with protection from apoptosis (Kaplan and Miller, 2000), we designed experiments to determine whether the anti-apoptotic activity of ICP10 PK depends on activation of this pathway and its downstream targets in hippocampal cultures. The results of these experiments were presented in the Specific Aim 2 Section. They suggest that, in hippocampal neurons, the anti-apoptotic activity of HSV-2 involves activation of the Raf/MEK/ERK pathway, and appears to be mediated by ICP10 PK present in the virion tegument (Smith *et al.*, 1994). Moreover, in HSV-2-infected cells, the levels of CREB (a downstream target of Ras/Raf/MEK/ERK pathway) and Bcl-2 (whose transcription is CREB-dependent) were stabilized, and Bag-1 was specifically-induced by ICP10 PK. The following comments seem pertinent with respect to these findings.

In the first series of experiments, activation of the Raf/MEK/ERK pathway was analyzed by analysis of virus- or mock-infected cell extracts by immunoblotting with antibodies specific for phosphorylated (activated) ERK or total ERK. The levels of activated (phosphorylated) ERK 1, 2 (P-ERK 1, 2) were significantly higher in hippocampal cultures infected with HSV-2 than in mock-infected cultures, but this increase was not seen in cells infected with ICP10ΔPK or HSV-1. ERK activation was

inhibited by U0126, which is a MEK-specific inhibitor (Favata *et al.*, 1998), suggesting

that the ERK activation is MEK-dependent. U0126 treatment of HSV-2-infected cultures

caused a dose-dependent increase in the proportion of apoptotic cells, suggesting that

MEK/ERK activation is involved in the anti-apoptotic activity of HSV-2. The contribution

of the upstream activator of MEK, i.e. c-Raf, to the anti-apoptotic activity of HSV-2, was

studied by infecting hippocampal cultures that express the c-Raf dominant negative mutant

K375M (Soh *et al.*, 1999). The c-Raf-K375M-FLAG expression vector used in these

experiments, expresses the dominant negative c-Raf-1 (contains a K375→M point

mutation), which is kinase defective, autophosphorylates weakly and inhibits the activity

of its normal cellular counterpart (Morrison *et al.*, 1993; Soh *et al.*, 1999). It also contains

the FLAG marker which allows for determination of transfection efficiency by

immunostaining with specific antibodies. At 24 hrs post-transfection, approximately 35-40

% of hippocampal cells expressed c-Raf K375 M, as determined by immunostaining with

antibody specific for the FLAG marker. Staining was minimal in non-transfected cultures.

When examined by TUNEL, apoptosis levels were increased by approximately 50 % in

cultures infected with HSV-2 in the presence of c-Raf-1 K375M, compared to the levels

obtained in cultures infected but not transfected. At this time it is not known whether

apoptosis occurred only in cells expressing c-Raf-1 K375M or in neighboring cells as well

(by a mechanism which may involve glutamate release from the dying, c-Raf-1 K375M-

expressing cells). The difference in transfection efficiency versus apoptosis levels suggests

that this may occur, but co-localization studies (FLAG expression-TUNEL) are needed to clarify this issue. Alternatively, this difference may be due to experimental/statistical error and/or distinct sensitivities of the different assays. By contrast, expression of c-Raf-1 K375M did not affect the survival of mock-infected hippocampal neurons, suggesting that activation of Raf kinase and therefore the downstream components MEK/ERK is not required for the basal maintenance of these neurons but becomes essential in the protection against virus infection. Presumably, activation of the Raf/MEK/ERK pathway occurs upstream of caspase-3 and it involves the activation of Ras, since i) ICP10 PK activates Ras, and thereby Raf/MEK/ERK, in non-neuronal cells, and ii) activation of Ras effector pathways protects neurons from apoptosis induced by various stimuli (Xia *et al.*, 1995; Erhardt *et al.*, 1999; Kaplan and Miller, 2000). However, we do not know the exact kinetics of Raf/MEK/ERK activation as they relate to inhibition of apoptosis. Indeed, activation of MEK/ERK was observed as early as 30 min post-infection, while apoptosis was only studied at 24 hrs post-infection, when MEK/ERK were no longer activated.

In non-neuronal cells, MEK/ERK are activated at 2 to 3 hrs post-infection by newly synthesized ICP10 PK (Smith *et al.*, 2000). By contrast, in HSV-2-infected hippocampal neurons, activation occurred as early as 30 min post-infection and did not require *de novo* viral protein synthesis, suggesting that it may be mediated by the virion ICP10 PK (Smith and Aurelian, 1997), which is released into the cytoplasm upon cellular penetration and virion uncoating. Alternatively, MEK/ERK activation could be a non-specific response to the mechanics of infection or due to growth factors or cytokines present in the inoculum. This is particularly significant, since an HSV receptor is a member of the TNFR family, which, upon ligand binding, can generate a signal that regulates NF-κB and AP-1 activation (Marsters *et al.*, 1997).

We favor the former interpretation because: i) MEK/ERK were not activated by ICP10ΔPK or HSV-1, the adsorption and penetration of which are identical to those of the wild type HSV-2 (Smith *et al.*, 1998), ii) for all viruses, inoculum contained similar serum %, and iii) MEK/ERK are activated within 30 min post-infection, prior to detectable *de novo* viral protein synthesis. This assumption is also supported by the observations that: i) ICP10 was not seen in cells infected with antibody-neutralized HSV-2, which can attach to but does not penetrate the cells (Highlander *et al.*, 1987; Johnson *et al.*, 1990), and the levels of ERK 1, 2 were not increased in these cells, and ii) MEK/ERK were activated equally well by HSV-2 and UV-inactivated HSV-2, which can penetrate the cells but is defective in protein synthesis (Purifoy and Powell, 1977). However, because anti-HSV-2 antibody blocks ERK 1,2 activation, the possibility cannot be excluded that virus binding *per se* also contributes to ERK activation. Thus, it remains to be determined whether and how binding *per se* contributes to ERK activation and whether the virion ICP10 PK is involved in MEK/ERK activation in any non-dividing cells or only in hippocampal cells.

Consistent with previous reports which implicate the PI3-K/Akt pathway in the neuronal survival (Downward, 1998; Kaplan and Miller, 2000), we found that the proportion of apoptotic cells in uninfected cultures was increased by treatment with LY294002, which specifically inhibits PI3-K-the upstream activator of Akt (Vlahos *et al.*, 1994). Similar results were also obtained with wortmannin (100 to 200 nM), which is another PI3-K inhibitor (Yano *et al.*, 1998). However, these inhibitors had a minimal, if any, effect on apoptosis in HSV-2-infected hippocampal cultures, suggesting that infection uncouples the survival of these cells from PI3-K activity. Presumably, uncoupling is due to the activation of the MEK/ERK pathway. Significantly, at 30 min post-infection Akt was activated/phosphorylated equally well in HSV-2 and ICP10ΔPK-infected cells, compared to the levels observed in mock-infected cultures, suggesting that a viral gene (other than ICP10 PK) present in the virion tegument, is responsible for this effect. However, by 24 hrs post-infection it becomes evident that the P-Akt levels were upregulated in HSV-2- and ICP10ΔRR-infected cells, but only minimally in HSV-1- and ICP10ΔPK-infected cells, suggesting that ICP10 PK contributes to this effect. Since Akt is equally activated by HSV-2 and ICP10ΔPK at 30 min post-infection, the contribution of ICP10 PK to increased activation by 24 hrs p.i. may involve *de novo* phosphorylation/activation of Akt, and/or increased protection of P-Akt from proteolytic degradation by caspases (François and Grimes, 1999) in the course of infection with ICP10 PK-expressing viruses. Additional studies, involving correlation between the kinetics of caspase-3 and Akt activation are necessary to clarify these issues. Akt activation was PI3-K-dependent because treatment with LY294002 completely inhibited it in HSV-2-infected cultures. The possibility cannot be excluded that both pathways inhibit pro-apoptotic pathways that lead to cell death. Such a pathway is JNK/c-Jun, that can lead to increased transcription of pro-apoptotic genes (such as Bad, Bax and p53) and apoptosis (Kaplan and Miller, 2000). Phosphorylation of c-Jun transcription factor at Ser 63 and Ser 73 residues is associated with c-Jun activation and specific gene transcription, including c-Jun gene itself (Bossy-Wetzel *et al.*, 1997; Minden and Karin, 1997). At 24 hrs post-infection, activation of c-Jun was slightly higher in HSV-2- than mock-infected cells. Also, treatment with LY294002 for 24 hrs induced the c-Jun activation, in both mock- and HSV-2-infected cultures. Presumably, activation of JNK/c-Jun pro-apoptotic pathway in HSV-2 infected cultures (treated with LY294002 or non-treated) is counteracted by activation of Raf/MEK/ERK pathway, which leads to activation of transcription factors (e.g. CREB, through Rsk) and transcription of genes associated with survival (e.g. Bcl-2). Such interpretation is consistent with previous reports that indicated that activation of the MEK/ERK pathway interferes with apoptosis at the level of cytosolic caspase activation, downstream of the release of cytochrome c from mitochondria, without the contribution of PI3-K/Akt pathway (Erhardt *et al.*, 1999) by transcription-dependent and -independent mechanisms (Xia *et al.*, 1995). By contrast, activation of JNK/c-Jun pro-apoptotic pathway in non-infected hippocampal cultures treated with LY294002, occurs in the conditions in which MEK/ERK are not activated, and therefore leads to cell death undeterred. We cannot exclude the possibility that in HSV-2-infected cultures there is cross talk between Raf/MEK/ERK and PI3-K/Akt pathways, because: i) ICP10 PK activates MEK/ERK at 30 min post-infection, and Akt at 24 hrs post-infection, ii) both pathways may activate CREB [which is associated with cell survival (Kaplan and Miller, 2000)], and iii) both pathways may have transcription-independent effects on survival, such as phosphorylation/inactivation of the pro-apoptotic protein Bad (Wang *et al.*, 1996; Datta *et al.*, 1997; Shimamura *et al.*, 2000), which were not examined in this study.

The targets of survival or apoptotic signaling pathways are transcription factors that lead to specific gene transcription. For example, CREB activation by phosphorylation at Ser 133 residue occurs as a neuronal response to neurotrophins and is important for the survival of neurons in a number of different conditions (Finkbeiner *et al.*, 1997).

Moreover, CREB activation couples alterations in gene expression not only with survival

but also with neuronal activity, since it has been implicated in synaptic plasticity and

cognitive functions (Bailey *et al.*, 1996; Silva *et al.*, 1998). The ERKs play a critical role

in CREB phosphorylation/activation in the hippocampus by either triggering it (via the

intervening kinase RSK) or acting as an obligatory intermediate or conduit in PKA and

PKC activation of CREB (Roberson *et al.* 1999). Moreover, the CaMK pathway will most

likely recruit the ERK pathway in response to strong Ca^{2+}-mobilizing stimuli thus leading

to a more prolonged increase in CREB activation (Wu *et al.*, 2001).

To assess CREB activation, hippocampal neurons in culture were infected with

HSV-2 or ICP10ΔPK and analyzed at various times p.i. by immunoblotting with

antibodies specific for Ser 133 phosphorylated (activated) CREB or total CREB. After 30

min p.i., when CREB was equally activated by the two viruses, HSV-2 continued to

activate CREB in a stable manner up to 24 hrs p.i. (when CREB activation in ICP10ΔPK-

infected cells was below that seen in mock-infected cultures), suggesting that CREB

activation may be involved in the anti-apoptotic activity of ICP10 PK. Moreover, ICP10

PK protected CREB from proteolytic cleavage by caspases, as evidenced by the presence

of p30 CREB cleavage product (François *et al.*, 2000) in ICP10ΔPK- but not HSV-2-

infected cells. This is a significant finding in view of CREB function without activation by

phosphorylation. Thus, it has been reported that non-phosphorylated CREB has anti-

apoptotic activity and can suppress AP-1 (c-Jun/c-Fos heterodimer) transcriptional activity

by competing for the AP-1 binding site on target genes (Masquilier and Sassone-Corsi,

1992). Significantly, LY294002 treatment did not affect the HSV-2-induced CREB activation at 24 hrs post-infection, suggesting that the PI3-K/Akt is not involved in CREB activation. However, the possibility cannot be excluded that PI3-K/Akt may contribute to CREB phosphorylation seen in HSV-2- and ICP10ΔPK-infected cells at 30 min post-infection, since, at this time, both viruses activate Akt equally well. 24 hrs treatment with LY294002 of non-infected hippocampal neurons, induced decreased levels of P-CREB, compared to the basal levels seen in non-treated cells, suggesting that PI3-K/Akt is involved in CREB activation. Thus, CREB activation-as induced by different pathways may be the key factor responsible for the distinct fate of infected versus non-infected hippocampal neurons upon inhibition of PI3-K/Akt pathway.

Consistent with these interpretations, inhibition of α7 nAChR signaling [which occurs mostly by activation of PI3-K/Akt, but it also involves MEK/ERK (Dineley *et al.*, 2001; Kihara *et al.*, 2001)] by treatment with the α7 nAChR antagonist, α-BGTX, inhibited CREB activation in mock-, but not HSV-2-infected cultures. However, the exact contribution of each of these pathways to α7 nAChR signaling, is unclear. It is plausible that MEK/ERK signaling through α7 nAChR constitutes only a minor contribution to this pathway activation in HSV-2-infected cultures, because: i) CREB activation levels are not affected by α-BGTX treatment, and ii) α-BGTX induces only low levels of apoptosis (25-30%). In non-infected cultures, α-BGTX treatment induced similar levels (25-30%) of apoptosis but it also inhibited CREB activation, suggesting that the mechanism of cell death upon α-BGTX treatment may be distinct in mock- versus HSV-2-infected cells. Increased levels of c-Jun activation were seen upon treatment with α-BGTX in both mock-

and HSV-2-infected cultures, suggesting a role for c-Jun in this apoptosis paradigm. However, this activation was lower in α-BGTX- than LY294002-treated mock-infected cells, suggesting that α7 nAChR signaling is not solely responsible for PI3-K/Akt activation in these cells. Moreover, the increased phosphorylation of c-Jun at Ser 63 versus Ser 73 in non-infected cultures, upon treatment with α-BGTX, is unclear, because total c-Jun expression was increased [an indicator that c-Jun transcription factor is activated (Minden and Karin, 1997)], but previous reports (reviewed by Minden and Karin, 1997) suggested that phosphorylation of both residues is required for activation of c-Jun transcriptional activity. Ultimately, the specific combination of transcription factors (with CREB as a key factor) and the gene expression they regulate may determine the fate of non-infected versus HSV-2-infected neurons. Concomitant activation of pro- and anti-apoptotic signaling pathways may be farther downstream regulated by the specific gene expression they regulate. Such a regulation may involve the balance between pro- and anti-apoptotic Bcl-2 family members.

Bcl-2 family of proteins are important regulators of apoptosis, and their activities are induced/suppressed or mimicked by various viral genes. In infected hippocampal neurons, HSV-1 infection leads to suppression of Bcl-2 expression by 24 hrs p.i. (when more than 50 % of cells are apoptotic), whereas Bcl-2 is still expressed in HSV-2-infected cultures, suggesting that it may contribute to virus-induced survival activity. However, at this time, it is unclear whether ICP10 PK specifically contributes to Bcl-2 stabilization, or whether Bcl-2 gene transcription in HSV-2-infected cells is CREB-dependent, as previously reported (Kaplan and Miller, 2000). It is also unclear whether the decreased levels of Bcl-2 seen in HSV-1-infected cultures are due to its cleavage by active caspases (Grandgirard *et al.*, 1998), because we did not detect any Bcl-2 cleavage products. Even though stabilization of Bcl-2 levels can be correlated with the increased survival of HSV-2- versus HSV-1-infected hippocampal neurons, the contribution of Bcl-2 to this effect is not known, and more studies are necessary to determine whether Bcl-2 is indeed required for protection from virus-induced apoptosis.

Significantly, Bcl-2 prevents many, but not all, forms of apoptotic cell death (Tsujimoto, 1998) which suggests the existence of multiple independent intracellular mechanism of apoptosis. Alternatively, these additional pathways may involve proteins that differentially regulate Bcl-2 function (Boise *et al.*, 1993). Such a protein is Bag-1 which interacts physically and functionally with Bcl-2 to cooperatively interfere with the apoptotic cascade at the level of caspase activation (Takayama *et al.*, 1995; Bardelli *et al.*, 1996; Schulz *et al.*, 1997). It not only enhances anti-apoptotic activity of Bcl-2 but also inhibits apoptosis by itself (Takayama *et al.*, 1995). Moreover, Bag-1 (like Bcl-2) has been show to interact with Raf-1 and increase its kinase activity (Wang *et al.*, 1996). In order to determine whether HSV infection (and specifically the ICP10 PK gene) affects the Bag-1 expression levels, hippocampal cultures were infected with HSV-1, HSV-2, ICP10ΔPK or ICP10ΔRR, or mock-infected, and Bag-1 expression was determined by immunoblotting with specific antibody. Bag-1 expression was specifically induced by anti-apoptotic HSV-2 and ICP10ΔRR, but not pro-apoptotic HSV-1 and ICP10ΔPK, suggesting that Bag-1 is involved in ICP10 PK-mediated survival of infected hippocampal neurons. Notwithstanding, Bag-1 synthesis was mainly regulated by Raf/MEK/ERK and PI3-K/Akt pathways, and minimally by p38 MAPK, suggesting that Bag-1 is a common downstream effector of these signaling pathways. A common denominator for these pathways is their ability to phosphorylate/activate CREB. However, the contribution of p38 MAPK to Bag-1 expression is unclear, because, at the specific concentration used for this experiment (10 μM), the p38 MAPK inhibitor, SB203580, may also inhibit Akt, as previously described (Lali *et al.*, 2000). Moreover, the contribution of p38 MAPK to HSV-2-induced survival of hippocampal neurons and/or CREB activation was not studied. These results also suggest that Bag-1 is not solely responsible for the ICP10 PK-mediated survival, because inhibition of PI3-K/Akt pathway affects the Bag-1 levels but not the survival of HSV-2-infected hippocampal neurons. Moreover, as Bag-1 has been shown to bind and activate the c-Raf kinase (Wang *et al.*, 1996), data suggest that, during infection, the newly synthesized Bag-1 may participate in activation of Raf/MEK/ERK pathway by a feedback mechanism. Activation of Raf/MEK/ERK pathway may also contribute to anti-apoptosis by transcription-independent mechanisms, such as phosphorylation of Bcl-2 or Bad (Wang *et al.*, 1996; Datta *et al.*, 1997; Shimamura *et al.*, 2000), which were not examined in this study. Additional studies are required to determine the validity of these interpretations.

However, a critical role for the anti-apoptotic protein Bag-1 in protection from virus-induced apoptosis, is suggested by experiments involving ectopically expressed Bag-1. Transfection of hippocampal cultures with pJG4-5 mBag-1 expression vector and subsequent (24-48 hrs) infection with ICP10ΔPK or HSV-1 viruses, resulted in inhibition of PARP cleavage and DNA fragmentation, associated with virus-infection [as determined by staining with antibody specific for the cleaved fragment of PARP (p85PARP) and

TUNEL]. The difference in transfection efficiency (30-40 %) versus the increase in survival levels (approximately 50 %) may be due to experimental/statistical error and/or distinct sensitivities of the different assays (immunocytochemistry versus TUNEL). At this time it is unclear whether ectopically expressed- Bag-1 mediated this effect by itself or in cooperation with the endogenously expressed Bcl-2, or other co-factors. We cannot conclude from these data that the failure of ICP10ΔPK or HSV-1 to induce Bag-1 expression is solely responsible for their apoptotic activities, because other factors (such as activation of glutamate receptors and/or the JNK/c-Jun pathway) may also be involved. Understanding the mechanisms by which ICP10ΔPK and HSV-1 induce apoptosis may provide insights on the mechanisms by which ICP10 PK promotes survival, i.e. not only by activating anti-apoptotic pathways but also by inhibiting those that lead to apoptosis. Thus, the next section (Specific Aim 3) will examine the mechanism of cell death in ICP10ΔPK- or HSV-1-infected hippocampal cultures, and HSE-affected human brain.

Specific Aim 3. To examine the cell death mechanism in: a) ICP10ΔPK-infected hippocampal neurons, and b) HSV-1-infected hippocampal neurons in culture and HSE-affected human brain.

1.   **HSV-1 activates the JNK pathway in hippocampal neurons**. Previous reports showed that in non-neuronal cells, HSV-1 and HSV-2 preferentially take advantage of the preexisting signaling pathways such as Ras/MEK/ERK for HSV-2 or JNK and p38MAPK

for HSV-1, to induce cellular and/or viral gene expression to promote viral replication (McLean and Bachenheimer, 1999; Zachos *et al.*, 1999; Smith *et al.*, 2000; Zachos *et al.*, 2001). Moreover, as shown in previous experiments in this report, the Raf kinase and the MEK/ERK pathway are required for HSV-2 anti-apoptotic activity in hippocampal neurons. However, JNK activation (a known apoptotic stimulus in neurons) in HSV-infected hippocampal neurons has not been studied. Two series of experiments were done in order to examine whether JNK is also activated in HSV-1-infected primary hippocampal cultures. In the first series, extracts of cells infected with HSV-1 (m.o.i. 10) or mock infected with growth medium for 24 hrs were immunoblotted with antibody specific for activated (phosphorylated) JNK (P-JNK). Antibodies to the non-phosphorylated JNK and actin served as controls. Both the JNK and P-JNK antibodies recognize all 3 isotypes (JNK 1,2,3). The results are shown in Fig. 39 A. The JNK 1 isotype was seen in mock-infected cultures immunoblotted with the JNK antibody, but bands were not resolved when the same blot was stripped and assayed with the P-JNK antibody (Fig. 39 A, lane 1). By contrast, in HSV-1-infected cultures, immunoblotting with the JNK antibody identified all 3 JNK isotypes (JNK 1, 2, 3) and immunoblotting with the P-JNK antibody identified P-JNK 1,2 (Fig. 39 A, lane 2), indicating that HSV-1 induces expression of JNK 2, 3 and activation (phosphorylation) of JNK 1,2. The connection between induction of JNK 3 expression and apoptosis is not clear, since phosphorylated (activated) JNK3 species was not detected. This is not an artefact due to improper gel loading since the actin levels were virtually identical in both the mock and HSV-1-infected cell extracts (Fig. 39 A). Previous reports showed that only JNK1 and

JNK2 [but not JNK3-which is neuron-specific (Minden and Karin, 1997)] are able to bind and therefore activate their main transcription factor target c-Jun, which is involved in apoptosis (Gupta *et al.*, 1996).

In the second series of experiments, primary hippocampal cultures were infected with HSV-1 or HSV-2 (m.o.i. 10) or mock-infected with growth medium and examined for JNK activation by immunocytochemistry with the P-JNK antibody at 0.5 hrs p.i. The cultures were counterstained with hematoxylin and cells were counted in 5 randomly selected microscopic fields (containing at least 250 cells). Results are expressed as % P-JNK-positive cells $\pm$ SEM. Mock and HSV-2 infected cultures had a similar number of positive cells (17.5 ± 1.1 % and 13.4 ± 1.6 %, respectively, $p > 0.05$ by ANOVA) (Fig. 39 B). By contrast, there was a significant higher number of P-JNK-positive cells in HSV-1 infected cultures (30.7 ± 2.9 %, $p < 0.05$ vs HSV-2 or mock, by ANOVA). These data suggest that activation of JNK pathway and its downstream targets may be involved in HSV-1-induced apoptosis of hippocampal neurons.

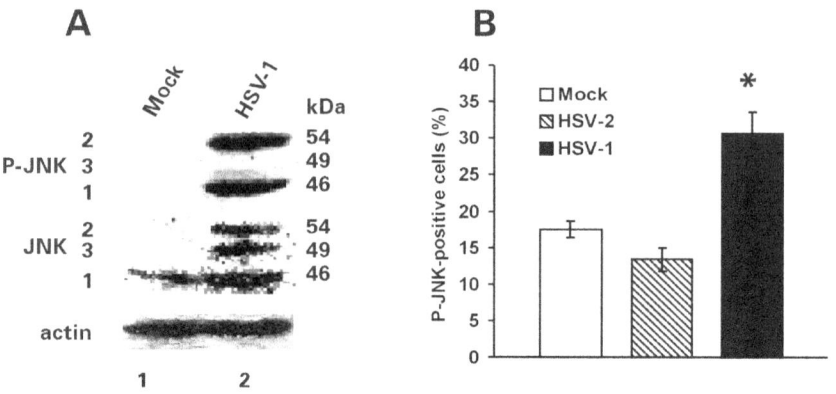

Fig. 39   HSV-1 activates JNK in hippocampal neurons. (A) Proteins from extracts of

hippocampal cells infected with HSV-1 (m.o.i. 10) or mock-infected for 24 hrs, were

resolved by SDS-PAGE, transferred to nitrocellulose membrane, and immunoblotted with

antibodies specific for P-JNK (upper panel). Blot was stripped and re-probed with

antibodies specific for JNK (middle panel) or actin (bottom panel, shown for loading

control).Lanes are identified by numbers at the bottom of the figure. **(B)** Primary cultures

of hippocampal neurons were either mock-infected or infected with HSV-2 or HSV-1

(m.o.i. 10) and stained with antibody specific for phosphorylated (active) JNK (P-JNK) at

0.5 hrs p.i. Cells were counterstained with hematoxylin and counted and results were

expressed as mean % P-JNK-positive cells $\pm$ SEM. Results shown are average of three

independent experiments. *, $p < 0.05$ vs. mock or HSV-2, by ANOVA.

2.   Virus-induced apoptosis of hippocampal neurons is associated with activation of specific transcription factors. Apoptosis has been shown to occur via a program that is dependent on macromolecular synthesis in some paradigms (Wyllie *et al.*, 1984), suggesting that this process involves the initiation of a set of genes that regulate events leading to or associated with cell death (Anderson *et al.*, 1995). Studies of neuronal death have suggested that the induction/activation of several transcription factors, such as c-Jun (Anderson *et al.*, 1995), ATF-2 (Yamada *et al.*, 1997), c-fos (Anderson *et al.*, 1994) or p53 (Sakhi *et al.*, 1994) is related to the initiation of apoptosis. To examine the association of these transcription factors with virus-induced apoptosis, several series of experiments were performed. In the first series of experiments, activation of c-Jun was analyzed in extracts of hippocampal cells infected with HSV-2, ICP10ΔPK, ICP10ΔRR, HSV-1 (m.o.i 10) or mock-infected (for 24 hrs) by immunoblotting with antibodies specific for the phosphorylated/activated c-Jun (P-Jun) or total c-Jun. The phosphorylation of c-Jun at serine residues (Ser) 63 and 73 (by JNK1 and JNK2, for example) has been previously shown to contribute to the induction of the transcriptional activity of c-jun (Behrens *et al.*, 1999). As shown in Fig. 40 A, HSV-1 (lane 5) induced significantly increased activation (phosphorylation at both Ser 63 and Ser 73) of c-Jun, when compared to mock- (lane 1) or HSV-2- (lane 2) induced levels. Increased c-Jun activation (phosphorylation) was also seen in ICP10ΔPK-infected cultures (Fig. 40 A, lane 3) compared to HSV-2 (Fig. 40 A, lane 2) or ICP10ΔRR (Fig. 40 A, lane 4), suggesting that ICP10 PK is responsible for inhibition of c-Jun activation. Similar kinetics were observed at 0.5 hrs p.i. (data not shown). To examine whether the HSV-1 counterpart of ICP10 PK (namely ICP6 PK) has a similar effect on c-Jun activation, primary hippocampal cultures were either mock-infected or infected with HSV-1 and two HSV-1 mutants: ICP6Δ (which has more than 90 % of the ICP6 gene coding sequence deleted) and *hr*R3 (which retains only 38 % of the N-terminus domain of ICP6 and has the RR domain replaced with *LacZ*) (Goldstein and Weller, 1987; Goldstein and Weller, 1988). At 24 hrs p.i. immunoblotting was done as described in Fig. 40 A. As seen in Fig. 40 B, only a minimal decrease in the levels of P-Jun (Ser63 or Ser73) were seen in ICP6Δ (Fig. 40 B, lane 3), or *hr*R3 (Fig. 40 B, lane 4)-infected cultures, compared to the levels in HSV-1-infected cultures (Fig. 40 B, lane 2), suggesting that ICP6 does not regulate c-Jun activation.

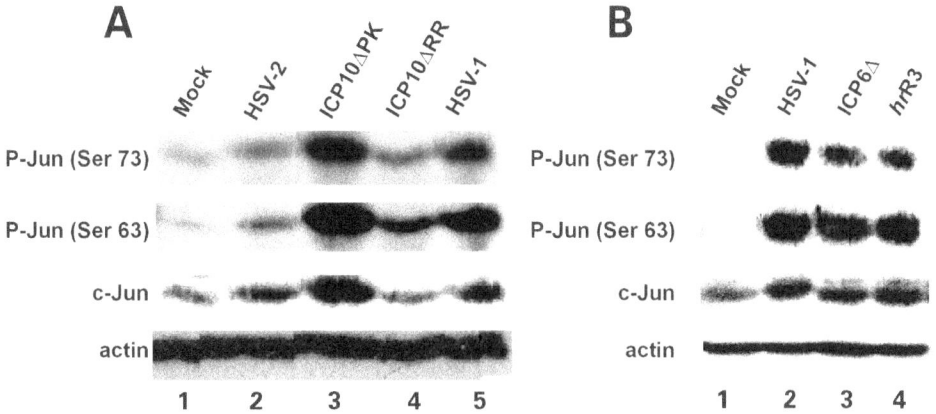

Fig. 40 ICP10 PK (but not ICP6PK) inhibits c-Jun activation. (A) Proteins from

extracts of hippocampal cells infected with HSV-1, HSV-2, ICP10ΔPK, ICP10ΔRR or

mock-infected for 24 hrs, were resolved by SDS-PAGE, transferred to nitrocellulose

membrane, and immunoblotted with antibodies specific for c-Jun phosphorylated (P-Jun)

at Ser 73 (upper panel) or Ser 63 (second from the top panel), c-Jun (recognizes c-Jun

independent of its state of phosphorylation) (third panel) and actin (bottom panel, for

loading control). Lanes are identified by numbers at the bottom of the panel. **(B)** Proteins

from extracts of hippocampal cells infected with HSV-1, ICP6Δ, *hr*R3 or mock-infected

for 24 hrs, were resolved by SDS-PAGE, transferred to nitrocellulose membrane, and

immunoblotted with antibodies specific for c-Jun phosphorylated (P-Jun) at Ser 73 (upper

panel) or Ser 63 (second from the top panel), c-Jun (recognizes c-Jun independent of its

state of phosphorylation) (third panel) and actin (bottom panel, for loading control). Lanes

are identified by numbers at the bottom of the panel.

The major regulators of the c-Jun promoter are ATF-2 and c-Jun (van Dam *et al.*, 1995).

The transcription factor ATF-2 (also called CRE-BP) binds to both AP-1 and CRE DNA

response elements, and is a target of both JNK and p38 MAPK signaling pathways (Gupta

et al., 1995; van Dam *et al.*, 1995). Cellular stress is known to activate ATF-2 by

phosphorylation of threonine residues 69 and 71 (Thr69/71) and mutations of these sites

results in the loss of stress-induced transcription (Gupta *et al.*, 1995; van Dam *et al.*,

1995). To examine the activation of ATF-2 in hippocampal neurons, cultures were

infected with HSV-2, ICP10ΔPK, or mock-infected, for 0.5, 8 or 24 hrs, and

immunoblotted with antibody that detects ATF-2 only when activated by dual

phosphorylation at Thr69/71 (P-ATF-2). As seen in Fig. 41 A, there was a basal activation

of ATF-2 in mock-infected cultures (lane 1) as evidenced by the two bands at 70-80 kDa,

consistent with previous reports that showed the presence of ATF-2 in hippocampus

(Yamada et al., 1997). By contrast, there was a significant reduction at 0.5 hrs p.i. in the

levels of P-ATF-2 in HSV-2-infected neurons (Fig. 41 A, lane 2), compared to those seen

in ICP10ΔPK-infected cells at the same time (Fig. 41 A, lane 3). By 8 hrs p.i. P-ATF

levels are almost similar in HSV-2 (Fig. 41 A, lane 4) and ICP10ΔPK-(Fig. 41 A, lane 5)

infected cells, and disappear at 24 hrs p.i. (Fig. 41 A, lane 6 and 7, for HSV-2 and

ICP10ΔPK, respectively). A similar kinetics was followed by c-fos protein levels, as

determined by immunoblotting with antibody specific for this protein (Fig. 41 A, middle

panel). The c-fos gene codes for a nuclear protein that dimerizes with the Jun family of

proteins to form the AP-1 transcription factor complex. As a member of the AP-1

complex, the c-fos protein has been implicated as a key molecule in cell proliferation, differentiation and transformation (Preston *et al.*, 1996); in addition, c-fos has been

previously associated with apoptotic death induced by anti-proliferative conditions

(Preston *et al.*, 1996) and neuronal injury (Anderson *et al.*, 1994). These data collectively

suggest that there may be many combinatorial associations between AP-1 (Jun/fos) and

ATF families (yielding different heterodimeric transcription factors with different DNA

binding specificities), that ultimately lead to specific gene transcription associated with

virus-induced apoptosis in hippocampal cultures. Notwithstanding, other transcription

factors, such as p53 (also a target of the pro-apoptotic JNK pathway), may also be

involved. To examine this possibility, primary cultures of hippocampal neurons were

infected with HSV-2, ICP10ΔPK, ICP10ΔRR, HSV-1 (m.o.i. 10) or mock-infected, and

the p53 protein levels were analyzed by immunoblotting with a specific antibody at 0.5

and 24 hrs p.i. As seen in Fig. 41 B, all four viruses induced increased expression of p53 at

0.5 hrs p.i., with higher levels seen in ICP10ΔPK (Fig. 41 B, lane 3) and HSV-1 (Fig. 41

B, lane 5), than HSV-2 (Fig. 41 B, lane 2) or ICP10ΔRR (Fig. 41 B, lane 4). This was not

due to improper gel loading because actin levels were identical in all samples (Fig. 41 B,

lower panel). p53 expression was below detectable levels in mock-infected cells at 0.5 hrs

p.i. and in mock- or virus-infected cells at 24 hrs p.i. (data not shown). Interestingly, the

p53 protein seemed to migrate slower in HSV-1-infected cell extracts, suggesting that it

may be post-translationally modified (phosphorylated). However, this pattern of migration

was not seen in ICP10ΔPK-infected cell extracts, which makes the association of p53

phosphorylation with apoptosis, difficult to interpret, unless HSV-1 and ICP10ΔPK

mutant induce apoptosis by different mechanisms. These data suggest that p53 expression is induced by HSV infection of hippocampal neurons, but its contribution to apoptosis is

unclea   r.

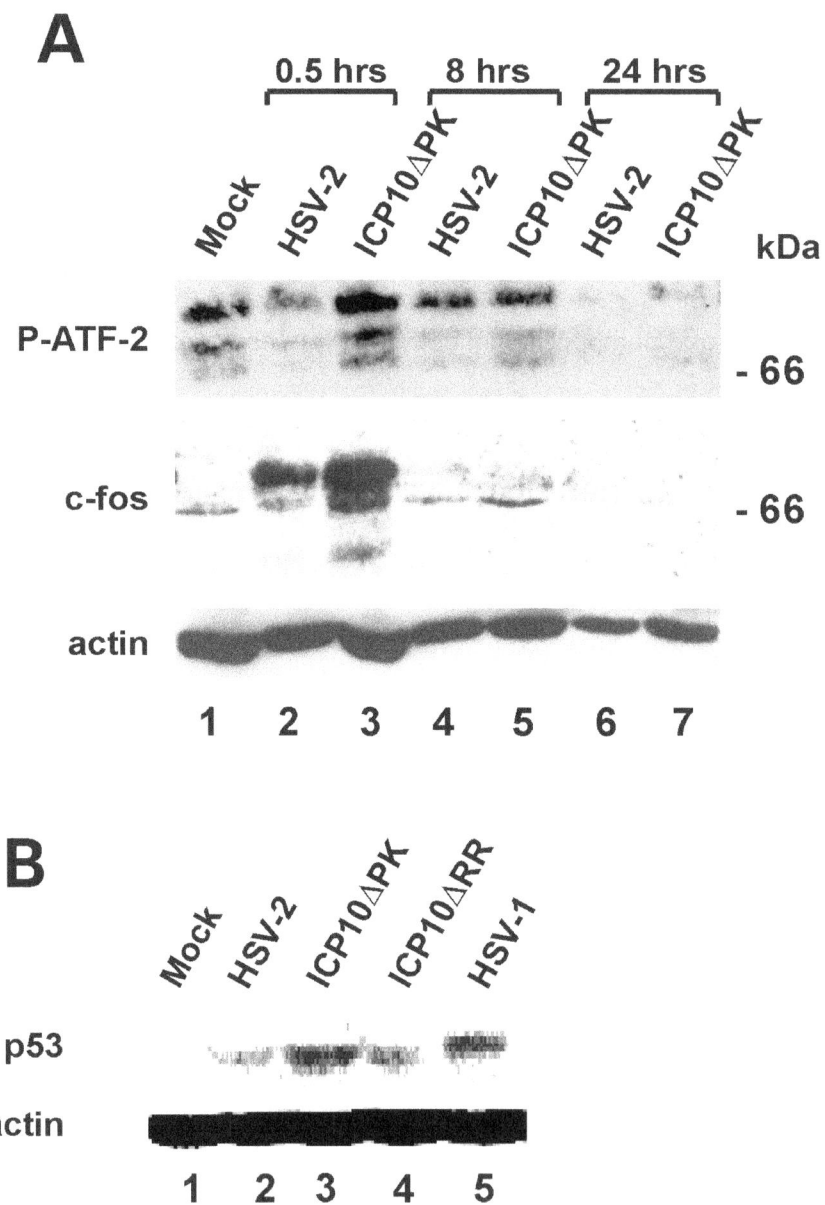

Fig. 41 Activation of apoptosis-associated transcription factors in virus-infected

hippocampal neurons. (A) Proteins from extracts of hippocampal cells infected with

HSV-2, ICP10ΔPK, or mock-infected for 0.5, 8 or 24 hrs, were resolved by SDS-PAGE, transferred to nitrocellulose membrane, and immunoblotted with antibodies specific for dually phosphorylated ATF-2 (P-ATF-2) (upper panel). The blot was stripped twice and reprobed with antibodies specific for c-fos (middle panel) and actin (bottom panel, for loading control). **(B)** Proteins from extracts of hippocampal cells infected with HSV-2, ICP10ΔPK, ICP10ΔRR, HSV-1, or mock-infected for 0.5, were resolved by SDS-PAGE, transferred to nitrocellulose membrane, and immunoblotted with antibodies specific for p53 (upper panel). The blot was stripped twice and reprobed with antibody specific for and actin (bottom panel, for loading control) or non-immune sera which was negative (data not shown).

3.   Virus-induced apoptosis associates with Bad expression in hippocampal neurons. A target of JNK/c-Jun-mediated gene transcription is the pro-apoptotic member of the Bcl-2 family, Bad (Bossy-Wetzel *et al.*, 1997). The Bad protein was first isolated on the basis of its interaction with Bcl-2 (Yang *et al.*, 1995). Bad is a pro-apoptotic member of this family that heterodimerizes with anti-apoptotic proteins such as Bcl-2 and Bcl-xL, promoting cell death (Yang *et al.*, 1995; Kelekar *et al.*, 1997). Induction of Bad expression is involved in glutamate induced apoptosis (Wang *et al.*, 1999) and neurodegenerative disorders (Kitamura *et al.*, 1998). In virus-infected hippocampal cultures, Bad expression was induced in cells infected with the apoptotic viruses HSV-1 and ICP10ΔPK at 0.5 hrs p.i. (Fig. 42, lanes 5 and 3 respectively). By contrast, Bad expression was below detectable

levels in mock- (Fig. 42, lane 1), HSV-2- (Fig. 42, lane 2) or ICP10ΔRR- (Fig. 42, lane 4)-

infected cultures. These data suggest that HSV-1 and ICP10ΔPK may activate apoptosis

by differentially regulating the levels of death-preventing versus death-inducing members

of the Bcl-2   family.

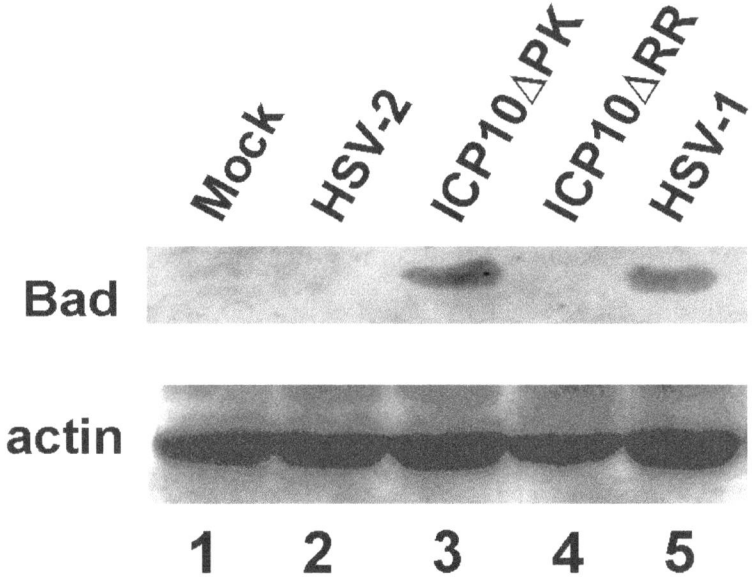

Fig. 42   Pro-apoptotic viruses induce Bad expression. Proteins from extracts of

hippocampal cells either mock-infected or infected with HSV-2, ICP10ΔPK, ICP10ΔRR

or HSV-1 (m.o.i. 10), or mock-infected, for 0.5 hrs, were resolved by SDS-PAGE,

transferred to nitrocellulose membrane, and immunoblotted with antibodies specific for

Bad (upper panel) or actin (bottom panel). Lanes are identified by numbers at the bottom

of the panel.

4.   **Apoptosis induced by ICP10ΔPK requires *de novo* viral protein synthesis.** In order to determine the contribution of viral proteins to the apoptotic activity of ICP10ΔPK, hippocampal cultures were infected with UV-inactivated ICP10ΔPK. Cultures infected with UV-inactivated HSV-2 were studied in parallel. UV inactivation of virions was as previously described (Purifoy and Powell, 1977). Analysis of apoptosis was done by TUNEL at 24 hrs p.i. and the results were expressed as % apoptotic cells $\pm$ SEM. UV inactivation resulted in a significant reduction in the levels of apoptosis triggered by ICP10ΔPK (from 68 $\pm$3.9 to 16.7 $\pm$ 4.1 %) (Fig. 43 A). Apoptosis was not completely abolished, most likely due to the fact that UV inactivation was not absolute but caused only a significant reduction in titer (of approximately 4-log, as determined by plaque assay). By contrast, the HSV-2 anti-apoptotic activity remained similar to that observed for untreated virus (3.7 $\pm$ 1.1 % for UV-inactivated versus 5.7 $\pm$ 0.5% for non-inactivated) (Fig. 43 A) suggesting that the HSV-2-anti-apoptotic activity does not require *de novo* viral protein synthesis and the anti-apoptotic gene is present in the virions. Since UV treatment damages the viral DNA and precludes the synthesis of viral proteins, it can be concluded that maximal apoptotic activity of ICP10ΔPK requires *de novo* viral protein synthesis.

Expression of viral proteins is coordinately regulated and sequentially ordered (IE, E and L proteins) (Roizman and Sears, 1996). To analyze the specific class of viral proteins involved in triggering apoptosis, a sequential combination of cycloheximide (CHX) and actinomycin D (Act D) (Honess and Roizman, 1974) was used. In this experiment, hippocampal cultures were either mock infected or infected with HSV-2 and ICP10ΔPK in the presence of 20 μg/ml CHX (protein synthesis inhibitor) for a period of 6 hrs (1 hr absorption + 5 hrs p.i.). during this time, protein synthesis was blocked and only the transcripts of immediate-early (IE) proteins were synthesized without any viral translation. CHX was removed at 6 hrs p.i. and 10 μg/ml Act D (transcription inhibitor) was added for 18 hrs, allowing the translation of preexisting IE transcripts, but precluding the synthesis of early (E) and late (L) viral proteins. TUNEL was performed at 24 hrs p.i. and the results are presented in Fig. 43 B. Treatment with metabolic inhibitors induced increased apoptosis in mock-infected cells (from 5 ± 2 % to 21.7 ± 5.6 %) ($p<0.05$, by ANOVA), suggesting that protein synthesis is required for the basal maintenance of these cultures. Treatment with metabolic inhibitors did not change the low levels of apoptosis observed for HSV-2 infected cells (11.9 ± 4.2 % and 10.7 ± 1 % for treated and untreated, respectively) ($p>0.5$, by ANOVA), suggesting that protection from apoptosis does not require *de novo* E and L protein synthesis. These data are in contrast to those obtained in non-neuronal cells for the anti-apoptotic HSV-1 protein ICP27 which requires that the infection proceeds to the stage in which viral DNA synthesis takes place and both early and leaky -late (γ_1) proteins accumulate (Aubert *et al.*, 2001). Treatment with metabolic inhibitors decreased the apoptosis levels in ICP10ΔPK infected cells from 73.9 ± 3.9% to 17.6 ± 2 % ($p<0.01$, by ANOVA), suggesting that maximal apoptotic activity of ICP10ΔPK requires synthesis of E and/or L viral proteins.

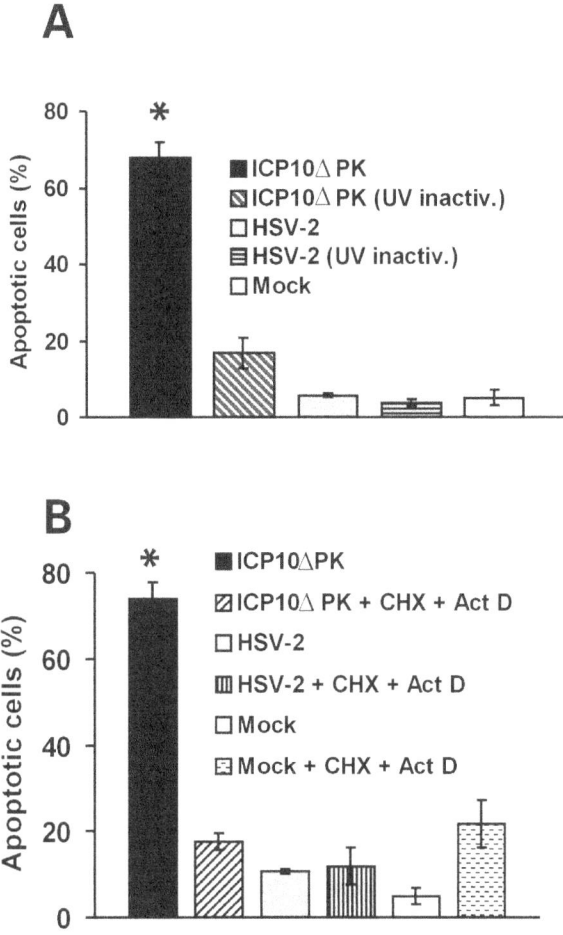

Fig. 43 **Apoptosis induced by ICP10ΔPK requires *de novo* viral protein synthesis.**

(A) Primary cultures of hippocampal neurons were infected with HSV-2 and ICP10ΔPK or

UV inactivated HSV-2 and ICP10ΔPK viruses for 24 hrs. Apoptosis was determined by

TUNEL. Results are expressed as % apoptotic cells $\pm$ SEM; *, p<0.01 vs HSV-2 or UV-

inactivated HSV-2, mock, and UV-inactivated ICP10ΔPK-infected cultures. **(B)** Primary

hippocampal cultures were either mock infected or infected with HSV-2 and ICP10ΔPK in

the presence or absence of 20 μg/ml cycloheximide (CHX) for 6 hrs (1 hr absorption + 5 hrs p.i.). CHX was removed at 6 hrs p.i. and 10 μg/ml Act D was added for 18 hrs. TUNEL was performed at 24 hrs p.i. Results are expressed as mean % apoptotic cells $\pm$ SEM; *, p<0.01 vs treated cultures infected with HSV-2, ICP10ΔPK, or mock, and untreated cultures mock-infected or infected with HSV-2.

5.   Virus-induced apoptosis involves activation of glutamate receptors. Activation of glutamate receptors has been shown to induce a trophic response in neurons (Bambrick *et al.*, 1995) and low concentrations of glutamate or glutamate agonists reportedly increase survival of postnatal cerebellar granule cells (Balasz *et al.*, 1988). However, high concentrations of glutamate trigger neuronal death. Excess glutamate overstimulates α-amio-3-hydroxy-5-methyl-4-isoxazole propionic acid (AMPA)-, kainate-, and N-methyl-D-aspartate (NMDA)-type glutamate receptors, resulting in an influx of cations (primarily Ca^{2+}, Na^+, and K^+) through channels gated by these receptors. The resulting elevation in intracellular Ca^{2+} activates phospholipases, oxidases, nitric oxide synthase, proteases and phosphatases and leads to lethal metabolic dysfunctions (Choi, 1988). Excitotoxicity has been implicated in the pathogenesis of lentivirus- and measles virus-induced CNS damage (Giulian *et al.*, 1990; Andersson *et al.*, 1991) and cytotoxicity associated with some HSV-1 vectors (Ho *et al.*, 1995).

To examine whether activation of glutamate receptors is involved in ICP10ΔPK-induced apoptosis, primary cultures were pre-treated with 10μM CNQX (antagonist of the AMPA/KA receptor) or 50 μM APV (antagonist of he NMDA-type receptor) for 30 min, then infected with ICP10ΔPK or HSV-2 (as control), or mock-infected, in the absence or presence of the inhibitors. Cells were analyzed by TUNEL at 24 hrs p.i. and results expressed as % apoptotic cells $\pm$ SEM. As shown in Fig. 44, neither APV or CNQX treatment induced apoptosis in mock-infected cultures [6.3 ± 2.2 % (not treated); 5.1 ± 0.5 % (CNQX-treated); 3 ± 0.5 % (APV-treated); $p > 0.05$, by ANOVA]. Similarly, APV did not affect significantly ($p > 0.05$, by ANOVA) the levels of apoptosis in ICP10ΔPK-infected cultures (62 ± 1.2 vs 57.3 ± 4.5). By contrast, CNQX treatment reduced significantly ($p < 0.01$, by ANOVA) the apoptosis levels, from 62 ± 1.2 % to 34.4 ± 1.8 %, suggesting that activation of AMPA/KA glutamate receptors may also contribute to the cell death in ICP10ΔPK-infected cultures. Unexpectedly, and by contrast to ICP10ΔPK-infected cultures, CNQX treatment triggered cell death not protection in HSV-2-infected cultures (11.5 ± 1.6 % in non-treated versus 30 ± 3.1 % in CNQX-treated cultures; $p < 0.05$, by ANOVA). Similarly, APV treatment also triggered cell death in HSV-2-infected cultures (11.5 ± 1.6 % in non-treated versus 59.1 ± 5 % in APV-treated cultures; $p < 0.01$, by ANOVA), suggesting that activation of glutamate receptors is required for survival of HSV-2-infected hippocampal neurons.

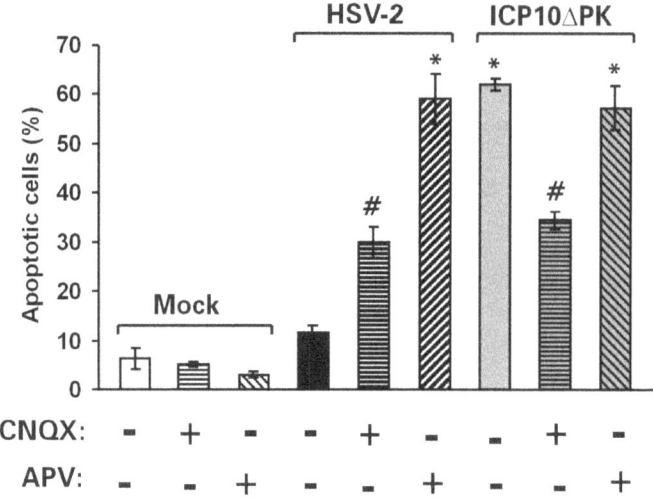

Fig. 44 ICP10ΔPK-induced apoptosis involves activation of glutamate receptors.

Primary hippocampal cultures were pre-treated with 10μM CNQX (inhibits the

AMPA/KA receptor) or 50 μM APV (inhibits the NMDA-type receptor) for 30 min, then

infected with ICP10ΔPK or HSV-2 (as control) at m.o.i. 10, or mock-infected, in the

absence or presence of the inhibitors. Cells were analyzed by TUNEL at 24 hrs p.i. and

results expressed as % apoptotic cells $\pm$ SEM. *, p<0.01, vs treated mock-infected and not

treated HSV-2- or mock-infected; #, p<0.05 vs treated mock-infected and not treated

HSV-2- or mock-infected by ANOVA.

6. **The HSV-1 counterpart of ICP10 PK (designated ICP6PK) does not inhibit PARP cleavage.** As shown in previous experiments in this report, HSV-1-infection triggers apoptosis in hippocampal cultures and, in contrast to ICP10 PK, the HSV-1 counterpart (ICP6 PK) does not inhibit c-Jun activation. In order to determine the role (if any) of ICP6 PK in the apoptotic pathway leading to PARP cleavage, hippocampal cultures were mock-infected or infected with HSV-1, HSV-2 (as control) and two HSV-1 mutants, ICP6Δ and *hr*R3 (respectively deleted in the ICP6 gene or the RR domain of ICP6). Infected cultures were stained with p85 PARP antibody (recognizes the enzymatically active 85 kDa PARP fragment) by immunoperoxidase method at 24 hrs p.i. As seen in Fig. 45, HSV-2 infection did not trigger PARP cleavage, and the % of p85PARP-positive cells in HSV-2- (9.9 ± 0.2 %) or mock- (4.2 ± 1.3 %) infected cultures were similar ($p>0.05$, by ANOVA). By contrast, increased PARP cleavage was seen in HSV-1 infected cultures (33.8 ± 2.2 %), similar to those in ICP6Δ (44.8 ± 9.1 %) and *hr*R3 (36.8 ± 2.7 %) infected cultures, suggesting that ICP6 PK does not have anti-apoptotic activity.

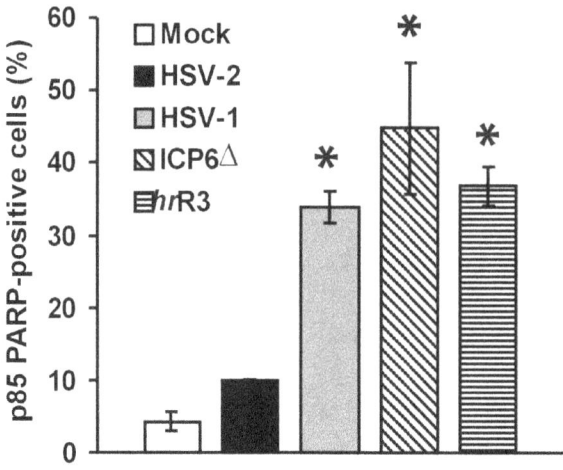

Fig. 45 **ICP6 PK does not inhibit PARP cleavage.** Primary cultures of hippocampal neurons were infected with HSV-1, ICP10Δ6, *hr*R3 or HSV-2 (as control), or mock-infected. At 24 hrs p.i. cultures were stained with antibody specific for the cleaved fragment of PARP (p85PARP) by the immunoperoxidase method. p85PARP-positive and -negative cells were counted and results expressed as % p85PARP-positive cells ± SEM. *, p<0.01, vs mock or HSV-2 infected cells, by ANOVA.

7.   Ectopically expressed ICP10 PK inhibits HSV-1- and ICP10ΔPK-induced apoptosis in hippocampal neurons. In order to determine whether the unique structural and functional features of ICP10 PK that distinguish it from its HSV-1 counterpart are sufficient to protect hippocampal neurons from HSV-1-induced apoptosis, cultures were transfected with pJW17 and pJHL15 expression vectors [encoding for ICP10 or the PK-negative p139[TM] mutant of ICP10, respectively (Chung *et al.*, 1989; Luo and Aurelian, 1992)] using the FuGene transfection reagent according to the manufacturer instructions. Expression of ICP10 and its mutant was seen in 35-40 % of cells as determined by immunocytochemistry with an ICP10 specific antibody (Aurelian *et al.*, 1989) at 48 hrs post-transfection (data not shown). Hippocampal cultures expressing ICP10 or p139[TM] were infected with HSV-1 or ICP10ΔPK (m.o.i.10) or mock infected and staining with p85 PARP antibody was done at 24 hrs p.i. The % p85 PARP-positive cells was similar in pJW17- (1.4 $\pm$ 0.6 %) or pJHL15- transfected (2 $\pm$ 0.4 %) and non-transfected (4.2 $\pm$ 1.3 %) mock-infected cultures ($p>0.05$, by ANOVA) (Fig. 46). By contrast, pJW17 transfection (ICP10 expression) induced a very significant reduction of PARP cleavage in ICP10ΔPK-infected cultures (3.5 $\pm$ 0.6 % versus 39.5 $\pm$ 2.5 %, $p<0.001$ by ANOVA). This reduction in PARP cleavage was ICP10PK-dependent because expression of the PK negative mutant of ICP10, p139[TM] (when transfection was done with pJHL15 plasmid) did not inhibit PARP cleavage (31.8 $\pm$ 4.5 % versus 39.5 $\pm$ 2.5 %, $p>0.05$ by ANOVA). Similarly, ICP10 inhibited PARP cleavage in HSV-1 infected cultures (5.1 $\pm$ 2.4 % versus 33.8 $\pm$ 2.2 %, $p <0.001$ by ANOVA), in a PK-dependent manner [because p139[TM] expression did not reduce the % of p85PARP-positive cells, 35.9 + 3.5 versus 33.8 $\pm$ 2.2

%, p <0.05 by ANOVA) (Fig. 46), suggesting that ICP10 PK protects from HSV-1-

induced apoptosis.

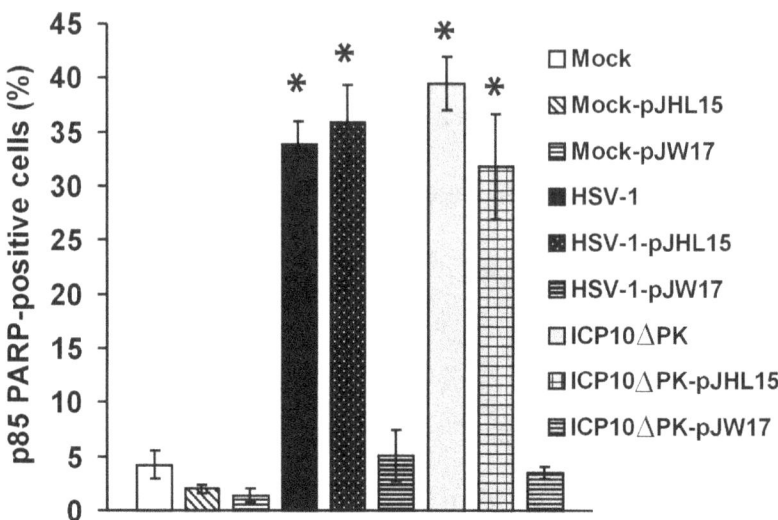

Fig. 46   Ectopically expressed ICP10 PK inhibits virus-induced PARP cleavage.

Primary cultures of hippocampal neurons were transfected with pJW17 or pJHL15

expression vectors (respectively coding for ICP10 or a PK-negative mutant of ICP10)

using the FuGene transfection reagent according to the manufacturer instructions. At 48

hrs p.i. transfected or non-transfected cultures were infected with HSV-1, ICP10ΔPK, or

mock-infected. At 24 hrs p.i. cultures were stained with antibody specific for the cleaved

fragment of PARP (p85PARP) by the immunoperoxidase method. p85PARP-positive and

207

-negative cells were counted and results expressed as % p85PARP-positive cells $\pm$ SEM.

*, p<0.001, vs mock (pJW17- or pJHL15-transfected or not transfected) and pJW17-transfected cultures infected with HSV-1 or ICP10ΔPK, by ANOVA.

8.   JNK is activated in human brain affected by HSV-1-induced encephalitis (HSE). In adults and older children, HSV-1 can cause a sporadic acute focal encephalitis (HSE) associated with severe destruction of temporal and frontal lobe structures including the limbic mesocortices as well as amygdala and hippocampus (Damasio and Van Hoesen, 1985; Whitley and Kimberlin, 1999). Since *in vivo* activation of JNK has been implicated in the pathogenesis of neurological disorders associated with apoptosis (Morishima *et al.*, 2001; Shoji *et al.*, 2001), and HSV-1 activates the pro-apoptotic JNK pathway in primary neuronal cultures (as described in the present studies), it seemed of interest to examine whether JNK is also activated *in vivo*, in the context of HSE-affected human brain. Eight HSE-typed brains (frontal cortex and hippocampus) were used for analysis. Two normal brains were used as controls (CTRL). To analyze JNK activation, immunohistochemistry with antibody specific for phosphorylated (activated) JNK (P-JNK) was done as described in Material and Methods. Significantly. Sections from 6 out of 8 HSE brains stained with P-JNK antibody. Fig. 47 A shows a representative micrograph of stained cells in the cerebral cortex of HSE-affected brain. No staining was seen in the neurons of control (normal) brains (Fig. 47 B) and the control experiments with non-immunized normal rabbit serum were negative (data not shown). These data suggest that JNK is activated in the context of   HSE.

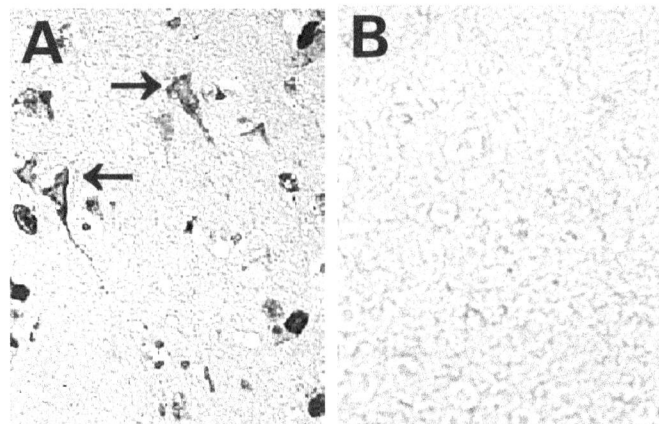

Fig. 47   **JNK is activated in HSE brains. (A)** Brain tissue samples obtained post-mortem from HSE patients were stained with antibody specific for phosphorylated (active) JNK (P-JNK) by the immunoperoxidase method. A positive signal was detected in glial and neuronal cells. Neurons are indicated by arrows. **(B)** Control (normal brains) were negative.

9.   **Apoptosis is present in HSE brains.** The presence of apoptosis in HSE brains was examined by immunohistochemistry with antibody specific for the large 17-20 kDa fragment of the active caspase-3 (Fig. 48 A). Staining was not seen in normal brains used as controls (CTRL) (Fig. 48 B, C). Upon counterstaining with hematoxylin, cells were counted and results were expressed as mean % caspase-3p20 positive cells $\pm$ SEM. The

results range from 0 to 14.5 $\pm$ 2.3 % caspase-3p20 positive cells, with 2 HSE brains

showing no signal (Fig. 48 C). The relative small percentage of cells positive for caspase-

3p20 may not be a good indicator of the real level of apoptosis but rather just a "snapshot"

in time, since apoptotic cells are rapidly cleared out *in vivo* (Johnson Webb *et al.*, 1997).

Interestingly, caspase-3 activation was seen in all HSE brains that showed JNK activation.

PARP cleavage, a crucial event in the commitment to undergo apoptosis (Johnson Webb

et al., 1997), was also seen in HSE brains (Fig. 48 D), as evidenced by

immunohistochemistry with antibody specific for the 85 kDa cleaved fragment of PARP

(p85 PARP). No specific stain was seen in control brain (Fig. 48 E) or when staining was

done with non-immunized normal rabbit serum (data not shown). Collectively, these

results suggest that caspase-3-dependent apoptosis may contribute to the pathological

manifestations of HSV-1-induced encephalitis (HSE) in a majority of cases.

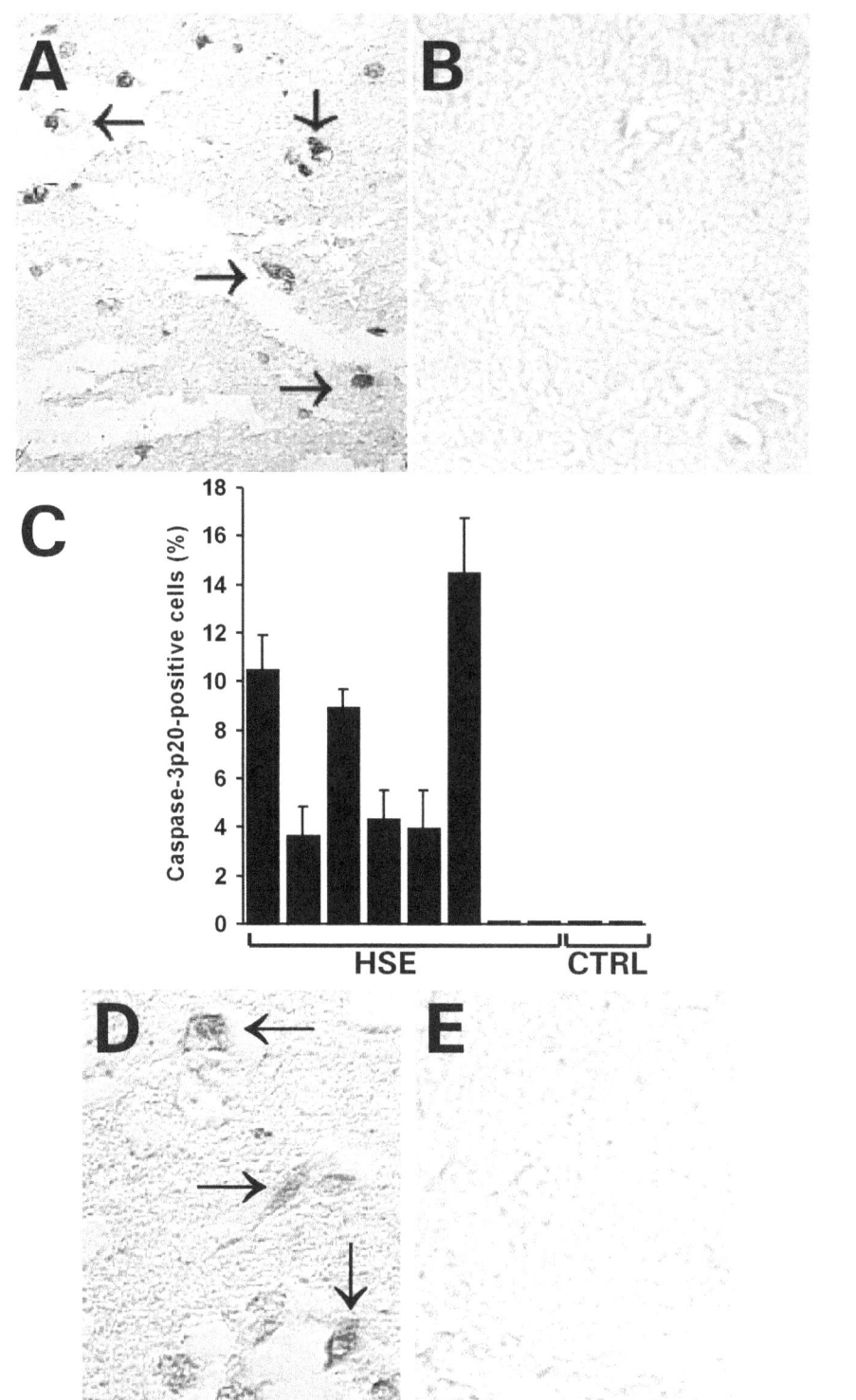

Fig. 48 **HSE is associated with caspase-3 activation and PARP cleavage.**

(A) Representative brain tissue sample obtained post-mortem from HSE patient, that was stained with antibody specific for cleaved (active) caspase-3 (caspase-3p20) by immunoperoxidase method. Positive signal was detected in glial, neuronal cells, and occasionally in cells of blood origin. Neurons are indicated by arrows. **(B)** Normal (control) brain showed no caspase-3p20 staining. **(C)** HSE brain samples processed as in (A) were counterstained with hematoxylin. Cells were counted and results were expressed as mean % caspase-3p20-positive cells $\pm$ SEM. **(D)** Representative brain tissue sample obtained post-mortem from HSE patient, that was stained with antibody specific for cleaved PARP (p85 PARP). Neurons are indicated by arrows. **(E)** Normal (control) brain showed no p85PARP staining.

10.   Discussion-Specific Aim 3

Both HSV-1 and HSV-2 have been shown to induce or protect cells from undergoing apoptosis. Induction of apoptosis occurs in a cell-type specific manner but the exact genes responsible have not yet been characterized. The mechanisms by which wt HSV or various HSV mutants (i.e. deleted in ICP27, US3, ICP4/US3 or US5) induce apoptosis are not yet known. However, since they induce cell death in a cell type-dependent manner, it is possible that the putative pro-apoptotic viral genes require specific cellular factors and/or conditions.

Consistent with the interpretation that ICP10 PK inhibits virus-induced apoptosis in cortical and hippocampal neurons, the HSV-2 mutant deleted in the PK domain of ICP10 (designated ICP10ΔPK) induced apoptosis (as determined by TUNEL and specific nuclear morphology) by activating effector caspases (including caspase-3) and promoting PARP cleavage (described in the Specific Aim 1 Section).

To determine the specific class of viral proteins involved in triggering apoptosis, we used a sequential combination of CHX and Act D (Honess and Roizman, 1974) or UV-inactivated ICP10ΔPK (deficient in viral protein synthesis). Apoptosis induced by ICP10ΔPK required *de novo* viral protein synthesis , because: i) apoptosis was not induced when infection was done with UV-inactivated ICP10ΔPK , and ii) apoptosis was inhibited when infection was done with ICP10ΔPK in the presence of metabolic inhibitors CHX and ActD. These data suggest that the pro-apoptotic viral gene responsible for this effect belong to the E and/or L class of viral proteins.

The hippocampal cell death triggered by ICP10ΔPK infection may partially involve activation of the ionotropic AMPA/KA receptor, since infection in the presence of the specific antagonist CNQX reduced the levels of apoptosis from about 60 % to about 35 %. By contrast, infection with ICP10ΔPK in the presence of the NMDA receptor specific antagonist APV did not affect the % apoptotic death, suggesting that NMDA receptor signaling is not involved in ICP10ΔPK-induced apoptosis. Significantly, even though overstimulation of both types of receptors (NMDA and AMPA/KA) was shown to induce the specific hallmarks of apoptosis (Simonian *et al.*, 1996; Ikonomidou *et al.*, 1999), differences between the NMDA and AMPA/KA-mediated demise of certain neuronal population were also described (Portera-Cailliau *et al.*, 1997). The AMPA/KA agonist kainic acid was reported to induce neuronal death through osmotic shock (Kiedrowski, 1998) or activation of NFκB that leads to p53 and c-myc induction and subsequent DNA fragmentation (Nakai *et al.*, 2000), whereas glutamate (agonist for NMDA) induced apoptosis due to mitochondrial collapse (Kiedrowski, 1998). The mechanism by which ICP10ΔPK induces activation of AMPA/KA receptors is not clear. Previous reports suggested that AMPA/KA receptor-mediated cell death is slow and requires many hours for cell damage to occur (Schubert and Piasecki, 2001). A potential mechanism for the ICP10ΔPK-induced apoptosis involves an initial population of neurons that is killed by virus directly, and another population that dies later by oxidative glutamate toxicity due to the activation of AMPA/KA receptors. According to this model, the first wave of cell death triggers cell damage and accumulation of glutamate in the culture medium due to the release of glutaminase from apoptotic cells, which converts the glutamine (present in the medium), to glutamate (Newcomb *et al.*, 1997). Because our hippocampal cultures lack glial cells, there is no effective way of removing glutamate, and thus the second wave of cell death occurs by overstimulation of AMPA/KA receptors (a process than can be inhibited by AMPA/KA antagonists, such as CNQX. What is the mechanism by which cells die in the putative first wave of ICP10ΔPK-induced cell death ? Presumably, it involves activation of pro-apoptotic signaling pathways, because ICP10ΔPK infection of hippocampal cultures is associated with: i) activation/phosphorylation of c-Jun at Ser 63/73 residues, both at 30 min (not shown) and 24 hrs post-infection, ii) activation/phosphorylation of ATF-2 at Thr 69/71 residues at 30 min post-infection, and up to 8 hr post-infection, iii) upregulation of c-fos expression at 30 min post-infection, and

iv) upregulation of p53 expression at 30 min post-infection, compared to the levels seen in mock- or HSV-2-infected cells. These transcription factors are known downstream targets of stress activated pathways such as JNK and p38 MAPK. However, at this time, it is unclear which of these pro-apoptotic pathways are activated by ICP10ΔPK. JNK pathway may activate c-Jun and ATF-2, which dimerize and bind to the c-Jun promoter, thereby stimulating the expression of the c-Jun gene (Minden and Karin, 1997). It may also mediate c-fos induction by phosphorylation of Elk-1, in the same manner as ERK. However, while induction of c-fos gene by growth factors is mediated primarily by ERK, JNK is most likely responsible for c-fos induction in response to cellular stress and cytokines (Minden and Karin, 1997). This is consistent with our observation that MEK/ERK are not activated by ICP10ΔPK infection. The contribution of p38 MAPK to ICP10ΔPK-induced activation of these specific transcription factors, cannot be excluded because p38 MAPK can also activate ATF-2 and mediate c-fos induction (Bonni *et al.*, 1995; Raingeaud *et al.*, 1996). p53 expression was induced by all viruses tested at 30 min post-infection, but it was upregulated in ICP10ΔPK- and HSV-1-infected cells. Whether this upregulation is an indicator of DNA damage associated with initiation of apoptosis, is unknown. However, the induction of p53 expression was associated with neuronal damage induced in the CNS by kainic acid (agonist for AMPA/KA receptor) (Sakhi *et al.*, 1994), which correlates with our results that suggest an AMPA/KA receptor-mediated component of ICP10ΔPK-induced apoptosis.

It seems likely that the specific signaling pathways activated or repressed during ICP10ΔPK infection of hippocampal neurons, are involved in virus induced apoptosis.

ICP10ΔPK does not activate the MEK/ERK pathway, even though p95 protein is expressed at 30 min p.i. similar to ICP10. However, ICP10ΔPK trigger some level of PI3-K/Akt pathway activation and CREB phosphorylation early in infection (30 min post-infection), but the significance of this effect is not clear and it does not seem to confer any protection against ICP10ΔPK-induced apoptosis. Presumably, inactivation of CREB by dephosphorylation or proteolysis late in infection (24 hrs p.i.) may contribute to ICP10ΔPK-induced apoptosis. Cleaved unphosphorylated CREB (p30) does not have anti-apoptotic activity (like p43, uncleaved unphosphorylated CREB) in that it cannot suppress the AP-1 (c-Jun/c-Fos heterodimer) transcriptional pro-apoptotic activity (Masquilier and Sassone-Corsi, 1992). Moreover, ICP10ΔPK fails to induce expression of the anti-apoptotic Bag-1 protein, but ectopic expression of Bag-1 was able to inhibit virus-induced apoptosis. By contrast, ICP10ΔPK infection triggers the expression of the pro-apoptotic Bcl-2 family member, Bad, implying that the balance between pro-and anti-apoptotic Bcl-2 family members is changed in favor of the pro-apoptotic proteins, thereby promoting cell death.

HSV-1 also caused apoptosis in hippocampal neurons in culture, as determined by TUNEL and specific nuclear morphology by activating effector caspases (including caspase-3) and promoting PARP cleavage. Similar to ICP10ΔPK, it did not activate the ERK pathway but it induced strong activation of the pro-apoptotic JNK pathway and its target c-Jun. The HSV-1counterpart of ICP10 PK (designated ICP6PK) did not have anti-apoptotic activity in hippocampal neurons. Thus, HSV-1 and its mutants, respectively deleted in the N-terminal (hrR3) or the ICP6 gene (ICP6Δ), induced similar levels of c-Jun activation and PARP cleavage. Presumably, HSV-1 also caused a disturbance in the balance of pro- and anti-apoptotic Bcl-2 family members (which is important in maintaining cell homeostasis), since it failed to induce Bag-1 expression but induced the expression of the pro-apoptotic protein Bad- that was previously reported to be induced by a JNK-dependent mechanism (Bossy-Wetzel *et al.*, 1997; Wan *et al.*, 2001)).

The significance of HSV-1-induced apoptosis in primary cultures of hippocampal neurons is best viewed in the context of virus pathogenesis in the CNS. In adults and older children, HSV-1 can cause a sporadic acute focal encephalitis (HSE) associated with severe destruction of temporal and frontal lobe structures which include the limbic mesocortices as well as amygdala and hippocampus (Damasio and Van Hoesen, 1985; Whitley and Kimberlin, 1999). In the experiments described in this section (Specific Aim 3) apoptosis (as determined by caspase-3 activation and PARP cleavage was present in human brain affected by HSE. Significantly, the pro-apoptotic JNK, which has been implicated in the pathogenesis of neurological disorders associated with apoptosis (Morishima *et al.*, 2001; Shoji *et al.*, 2001), was also found activated in HSE brains. The relatively low % of apoptotic cells present in the human brains affected by HSE may not be an accurate representation of the in vivo situation since apoptotic cells are rapidly cleared out in vivo by professional phagocytes or neighboring cells (Johnson Webb *et al.*, 1995). Necrosis, either due to the cytopathic effect of HSV-1 or inflammatory cytokines, is most likely another important factor in HSE etiology, and at this time, it is not clear whether apoptosis and necrosis follow a temporal pattern or occur as a continuum.

HSE is associated with high mortality (30 %) even in the presence of adequate antiviral treatment and may still leave survivors with severe neurological impairment (Whitley and Lakeman, 1995; Skoldenberg, 1996). As such, improvements in HSE therapy are still important research goals. Our findings implicate apoptosis in HSE pathogenesis and raise the possibility of optimizing HSE therapy and possibly improving the neurological outcome of HSE patients by combining anti-viral with anti-apoptotic therapeutic agents.

Significantly, ectopically expressed ICP10 rescued hippocampal cultures from ICP10ΔPK- or HSV-1-induced apoptosis by a PK-dependent mechanism, suggesting that ICP10 PK is the only viral gene responsible for the anti-apoptotic activity of HSV-2 in hippocampal neurons, and its unique structural and functional features distinguish it from its HSV-1 counterpart that has no neuroprotective activity.

The experiments designed to determine the mechanism of ICP10ΔPK- and HSV-1-induced apoptosis in hippocampal neurons also provided more information regarding the mechanism of survival in HSV-2-infected cells. Thus, when mock- and HSV-2-infected cultures were treated with glutamate receptor antagonists (as controls for ICP10ΔPK), we observed increased apoptosis in HSV-2-, but not mock-infected cells. CNQX and APV treatment triggered apoptosis in ~ 30 %, and 60 % of cells, respectively. This was a surprising result, and its interpretation would require additional experiments involving the kinetics of this death process. Presumably, this effect is a reflection of the role of glutamate receptors in survival. Previous reports showed that activation of glutamate receptors induces a trophic response in neurons (Bambrick *et al.*, 1995) and low concentrations of glutamate or glutamate agonists reportedly increase survival of postnatal cerebellar granule cells (Balasz *et al.*, 1988). The signaling pathways activated by AMPA/KA and in particular NMDA receptors, which are required for the survival of HSV-2-infected cells, may include ERK. This is consistent with previous reports that showed that stimulation of NMDA receptors in cultured neurons activates ERK and leads to CREB phosphorylation via calcium-dependent activation of NO synthase and NO production (Yun *et al.*, 1998). Therefore, NMDA receptor antagonists, such as APV, may act as potent inhibitors of MEK/ERK (which constitute the main survival pathway in HSV-2-, but not mock-infected neurons), leading to increased levels of apoptosis.

Another important conclusion that can be drawn from the experiments presented in this section (Specific Aim 3), is that ICP10 PK is involved in inhibition of the pro-apoptotic pathway JNK/c-Jun. Indeed, at 30 min post-infection, JNK was activated/phosphorylated in HSV-1-, but not HSV-2-infected cultures. At 24 hrs post-infection, c-Jun, the downstream target of JNK pathway (Minden and Karin, 1997), was strongly activated by HSV-1 and ICP10ΔPK, but only minimally by HSV-2 or ICP10ΔRR. Similar patterns of activation were seen at 30 min post-infection (data not shown). Activation of c-Jun (phosphorylation of both Ser 63 and Ser 73 residues) was associated with increased expression of c-Jun in ICP10ΔPK- and HSV-1-infected cells, consistent with previous reports that c-Jun activation regulates its own gene transcription (Angel and Karin, 1991). Interestingly, the ICP10ΔRR slightly higher levels of P-Jun (especially Ser 63) than HSV-2. It is unclear whether this effect is linked to the increased % of apoptotic cells (TUNEL-positive) seen in hippocampal cultures infected with ICP10ΔRR, compared to HSV-2-infected cells that suggested a contribution of the RR domain of ICP10 to protection from apoptosis. However, we did not see a significant increase in the c-Jun gene expression in ICP10ΔRR-infected cells, suggesting that the transcriptional activity of c-Jun is not fully activated solely by increased phosphorylation of Ser 63 residue, as previously reported (Minden and Karin, 1997).

What is the significance of the anti-apoptotic activity of ICP10 PK, which is exerted by activation of MEK/ERK/upregulation of Akt survival pathways and inhibition of JNK/c-Jun apoptotic pathway ? As discussed in Section I (Introduction), ERKs and CREB (downstream effector of MEK/ERK and PI3-K/Akt) have multiple beneficial roles in neurons, as they relate to neuron function and survival. Moreover, activation of JNK/c-Jun pathway was associated with neurological disorders that involve apoptosis, such as AD (Yamada *et al.*, 1997; Shoji *et al.*, 2001). Thus, the next series of experiments were designed to examine whether ICP10 PK can protects neurons from stimuli specific for neurological disorders.

Specific Aim 4. To determine whether ICP10 PK protects from apoptosis induced by stimuli specific for neurological disorders (i.e. growth factor withdrawal, oxidative stress or genetic defects that trigger apoptosis).

1.   ICP10 PK promotes survival of oxidatively stressed N2a cells that express mutant superoxide dismutase-1. Oxidative stress is associated with apoptosis in many neurological disorders, including ALS, AD, PD (Honig and Rosenberg, 2000), or DS (Sawa, 1999). ALS is characterized by degeneration of motor neurons. About 3-4 % of cases are associated with mutations (about 50 are known) in the gene encoding the free radical scavenging enzyme Cu, Zn superoxide dismutase1 (SOD1) (Honig and Rosenberg, 2000). The pro-apoptotic potential of mutant SOD1 is presumably due to an unstable conformation that leads to formation of toxic aggregates or to a gain in enzymatic function involving increased intracellular free radicals (Pasinelli *et al.*, 1998). To investigate whether ICP10 PK protects from apoptosis induced by mutant SOD1, mouse neuroblastoma cells that constitutively express wild type (wt) or mutant (G85R) SOD1 (Pasinelli *et al.*, 1998) were induced to differentiate to a neuronal phenotype by growth in serum-free medium, for at least 6 days, as previously described (Pasinelli *et al.*, 1998). Differentiated cells were transfected with pJW17 and pJHL15 expression vectors (that respectively express ICP10 or p139^{TM} -a PK negative mutant of ICP10). Expression of ICP10 and p139^{TM} was determined by immunoblotting with ICP10-specific antibody (recognizes amino acids 13-26 in both proteins) or normal rabbit serum (control), as previously described (Aurelian *et al.*, 1989; Luo and Aurelian, 1992). Approximately 35-

40 % of cells expressed the transfected proteins (data not shown). Differentiated but non-transfected cells were used as controls. At 24 hrs post-transfection, transfected or non-transfected cultures were treated with 100 µM xanthine and 10 mU/ml xanthine oxidase (X/XO) to generate superoxide anion and hydrogen peroxide, as previously described (Pasinelli *et al.*, 1998; Pasinelli *et al.*, 2000), and analyzed for viability using the CellTiter 96 Aqueous One Solution Cell Proliferation Assay (MTS assay) that measures enzymatic activity of functional mitochondria or TUNEL, at 4 or 6 hrs post-treatment. The results of the MTS assay (Fig.49 A) show that, upon X/XO treatment for 6 hrs, the survival of wt N2a cells was significantly (p<0.01, by ANOVA) higher than that of similarly treated G85R cells that express the mutant SOD1 (67.5 ± 1% versus $40.7 + 0.6$ %, respectively), suggesting that the presence of mutant SOD1 confers increased vulnerability to oxidative stress. By contrast, the survival of X/XO-treated G85R expressing the ICP10 protein was significantly (p<0.01, by ANOVA) higher (58.2 ± 0.7 % for G85R-pJW17 transfected versus $40.7 + 0.6$ % for G85R-non-transfected, respectively), suggesting that ICP10 protects cells expressing mutant SOD1 from oxidative stress. This protection was PK-dependent because the survival levels of G85R cells expressing the PK negative mutant p139^{TM} were similar (44.5 ± 0.2 %) (p>0.05, by ANOVA) to those of non-transfected G85R cells ($40.7 + 0.6$ %). In a second experiment G85R cells transfected (as above), were treated with X/XO for 4 hrs and analyzed by TUNEL (Fig. 49 B). Non-transfected G85R were used as controls. The % apoptotic cells was extremely low in non-treated G85R either transfected with pJW17 (4 ± 1.5 %) or pJHL15 (2 ± 0.4 %), or non-transfected (1 ± 0.3 %), suggesting that the presence of the mutant SOD1 is not sufficient

to cause apoptosis. However, when treated with X/XO, the % apoptotic cells increased significantly (p<0.01, by ANOVA) in both non-transfected (30.6 $\pm$ 4.5 %) and pJHL15-transfected cultures (26.5 $\pm$ 2 %), compared to both non-treated cells (1 $\pm$ 0.3 %). By contrast, the % apoptotic cells in G85R expressing ICP10 remained at very low levels even in the presence of X/XO treatment (4 $\pm$ 1.5 % versus 7.4 $\pm$ 1.4 %), suggesting that ICP10 PK protects cells that express mutant SOD1 from apoptotic death due to oxidative stress.

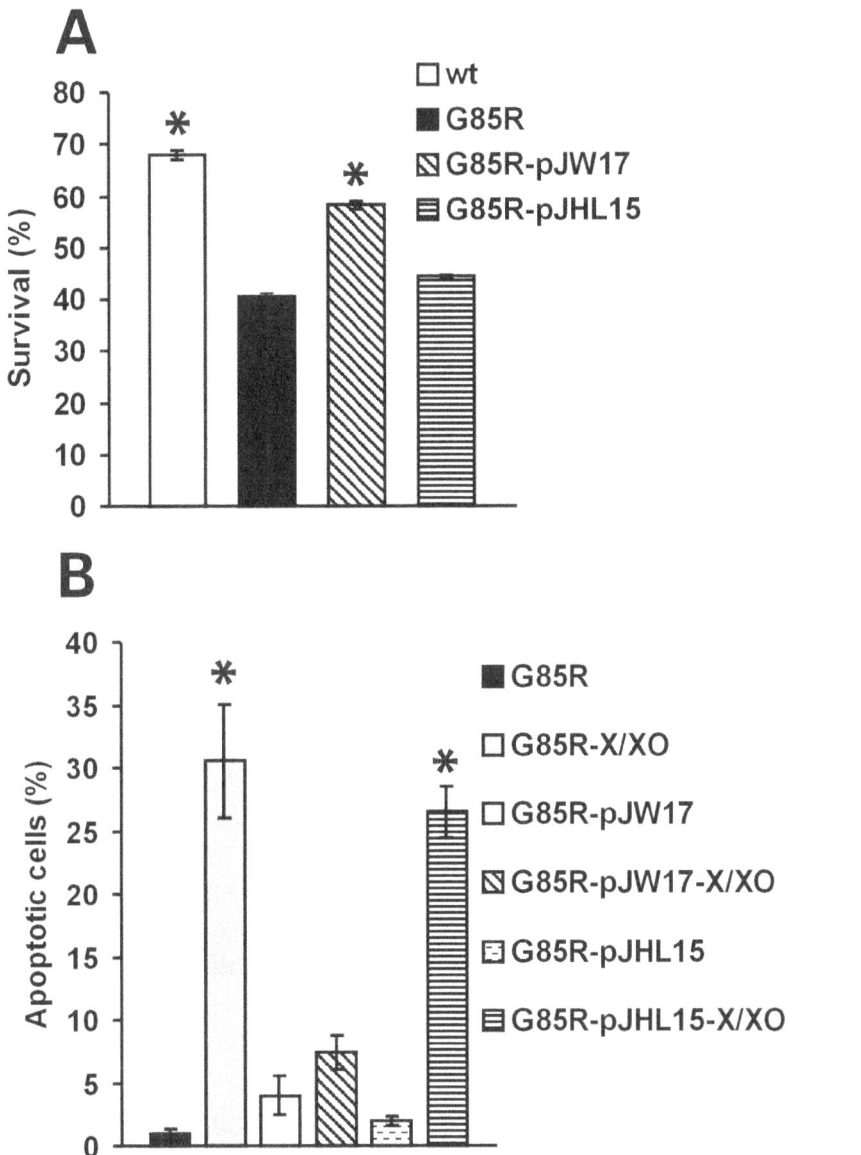

Fig. 49 ICP10 PK protects cells that express a mutant SOD1 (G85R) from apoptotic

death due to oxidative stress. (A) N2a cells that constitutively express wt SOD1

(designated wt) or mutant SOD1 (designated G85R) were neuronally differentiated by

growth in serum-free medium for at least 6 days. They were transfected with pJW17 and

pJHL15 expression vectors (that respectively express ICP10 or p139™ -a PK negative

mutant of ICP10). Differentiated but not transfected cells were used as controls. At 24 hrs post-transfection, transfected or non-transfected cultures were treated with 100 μM xanthine and 10 mU/ml xanthine oxidase (X/XO) for 6 hrs. Viability was determined by MTS assay and results are expressed as % viability at 0 hrs post-treatment $\pm$ SEM.*, $p<0.01$ vs G85R-X/XO and G85R-pJHL15, by ANOVA. **(B)** G85 R cells that were transfected and treated (X/XO) as above, were analyzed by TUNEL at 4 hrs post-treatment. Non-treated G85 R cells (transfected or not) were used as controls. Results are expressed as % apoptotic cells $\pm$ SEM. *, $p<0.01$ vs G85R, G85R-pJHL15 and G85R-pJW17, by ANOVA.

2.   ICP10 PK protects neuronal cells from death due to trophic factor deprivation. Since ICP10 PK activates the same survival pathway as the neurotrophins, it may be possible that it can substitute for trophic factor withdrawal and promote neuronal survival. In a first series of experiments to test this hypothesis, we used PC12 cells, which acquire properties of sympathetic neurons (e.g. neurite outgrowth, electrical excitability and expression of specific neuronal markers), when grown in NGF-containing medium, and die by apoptosis upon NGF withdrawal (Greene and Tischler, 1982). To test the protective activity of ICP10 PK in this system, PC12 cells were neuronally differentiated by growth (at least 12 days) in serum-free medium supplemented with 100 ng/ml NGF. They were transfected (using FuGene 6 Transfection Reagent) with vectors pJW17 or pJHL15, that respectively express ICP10 or the ICP10 PK negative mutant p139[TM] and cultured for 24 hrs to allow for transgene expression. At this time, the cells were washed, the medium was replaced with serum and NGF-free medium (0 hrs after NGF withdrawal) and the cells were cultured in this medium for an additional 48 hrs. They were examined daily for viability by MTS assay (using the CellTiter 96 Aqueous One Solution Cell Proliferation Assay, according to the manufacturer instructions). Results are expressed as % viable cells $\pm$ SEM relative to 0 hrs after NGF withdrawal.

Transgene expression was examined at 24 hrs post-transfection by immunoperoxidase staining with an ICP10-specific antibody (recognizes amino acids 13-26 in both proteins) or normal rabbit serum (control), as previously described (Aurelian *et al.*, 1989; Luo and Aurelian, 1992). The cultures were counterstained with Mayer's hematoxylin (Sigma), the stained cells were counted in 5 randomly chosen microscopic fields (containing at least 250 cells) and the average percentage of stained cells was calculated. The results are expressed as % positive cells $\pm$ SEM. Consistent with previous reports for non-neuronal cells transiently transfected with these vectors (Luo and Aurelian, 1992), staining was localized in the cytoplasm and both its intensity and the proportion of stained cells (25-35 %) were similar for pJW17 and pJHL15 (Fig. 50 B, C, respectively), suggesting that ICP10 and p139^{TM} are expressed equally well. Staining was not seen in control (non-transfected) PC12 cells (Fig 50 A), nor with normal rabbit serum (data not shown).

The kinetics of cell death in the non-transfected cultures were similar to those previously reported for this system (Pittman *et al.*, 1993), with a respective survival of 65.7 $\pm$ 2.7 % and 48.1 $\pm$ 1 % at 24 and 48hrs after NGF withdrawal). Similar kinetics were seen in pJHL15 transfected cultures (61.2 $\pm$ 1% and 43.1 $\pm$ 1.4 % at 24 and 48 hrs after NGF withdrawal, respectively). By contrast, the survival of pJW17- transfected cells

was 82.3 ± 3.9 % and 73.2 ± 3.3 % at 24 and 48 hrs after NGF removal ($p < 0.05$ vs

control and pJHL15 transfected cells, by ANOVA) (Fig. 51 D), suggesting that ICP10 PK

can compensate for the absence of NGF. Morphologically, both the non-transfected and

pJHL15-transfected cells showed degenerating cell bodies and "beading" of neurites

starting at approximately 24 hrs after NGF removal, while the pJW17-transfected cultures

were debris-free, with long neurites and cell bodies resembling those of NGF-treated PC12

cells (data not shown). Using these criteria, similar results were obtained in an

independent series of experiments with survivals (determined by counting

morphologically viable cells) of 70 ± 6.3% and 90 ± 4.7 % for non-transfected and

pJW17-transfected cells respectively, at 48 hrs after NGF withdrawal.

In a second series of experiments designed to examine the role of ICP10 PK in

neuronal survival, primary hippocampal cultures from embryonic (day 16) mice were

established on glass coverslips etched with a grid of 175 x 175 µm squares (CELLocate)

and grown (2 days) in MEM with B27 supplement which contains optimized

concentrations of neuron survival factors, as previously described (Bambrick and Krueger,

1999). At this time the cells were transfected with pJW17 or pJHL15, the medium was

replaced with MEM free of serum and growth factors (0 hrs). The cultures were then

maintained in this medium for 72 hrs. Non- transfected cultures maintained in medium

with B27 supplement (Eu+B27) or without the supplement (Eu-B27) served as controls.

Neuronal survival was determined daily by counting live cells (phase-dark bodies and fine

neurites) in seven randomly chosen squares (Bambrick and Krueger, 1999) and the results are expressed as % surviving cells $\pm$ SEM relative to 0 hrs. Neuronal identity was confirmed by staining with the neuron-specific antibody to class III ß tubulin (TuJ1) (Ferreira and Caceres, 1992). The % surviving cells were respectively 52.6 $\pm$ 7.2 and 56 $\pm$ 3.1 for Eu-B27 and pJHL15-transfected cells at 48 hrs and 34.3 $\pm$ 7.6 and 34.8 $\pm$ 3.8 % at 72 hrs. The viability of pJW17-transfected cells was significantly ($p<0.05$ by ANOVA) higher (80.6 $\pm$ 2.7 and 67.7 $\pm$ 2.4 % at 48 and 72 hrs, respectively) and similar to that of Eu+B27 cells (88.7 $\pm$ 3.1 and 80.6 $\pm$ 2.7 % at 48 and 72 hrs, respectively) (Fig. 50 D). The data suggest that ICP10 PK promotes survival of hippocampal neurons in the absence of growth factors.

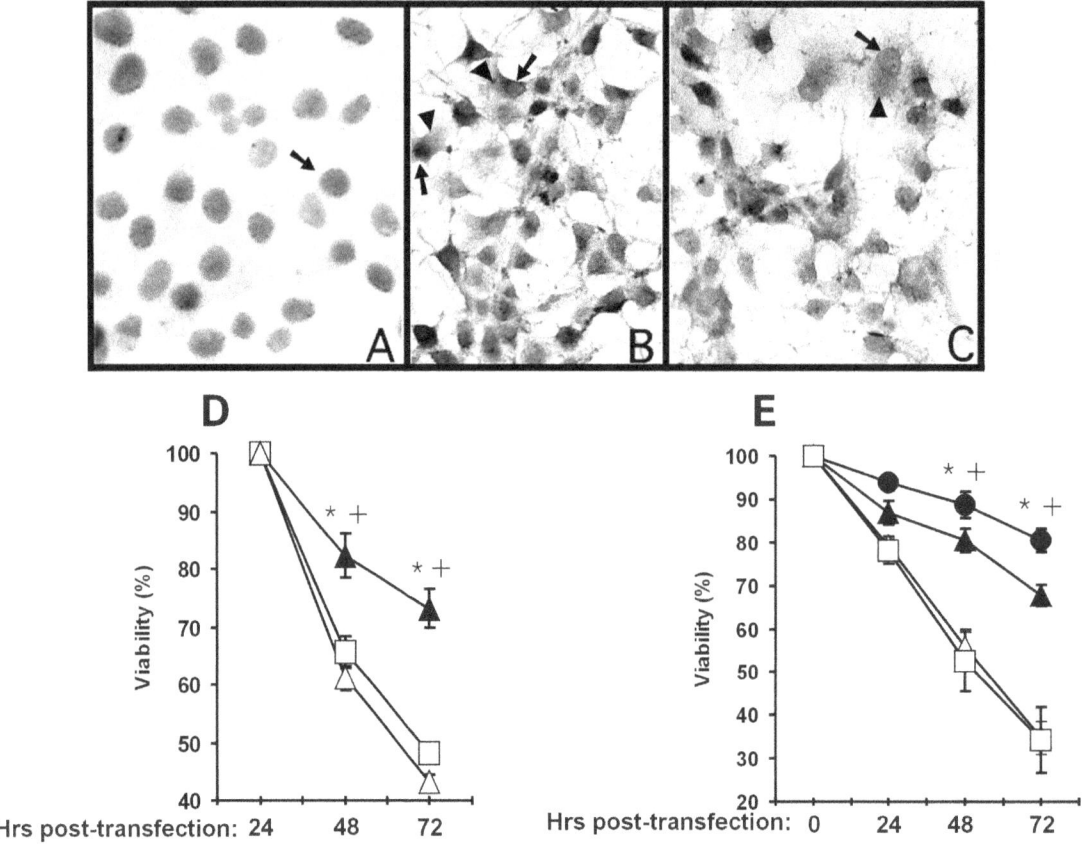

Fig. 50 **ICP10 PK protects neuronally differentiated PC 12 cells and primary**

hippocampal neurons from death due to growth factor withdrawal. PC12 cells

untransfected **(A)** or transfected with pJW17 [ICP10; **(B)**] or pJHL15 [p139™; **(C)**] were

stained with ICP10 specific antibody and counterstained with Mayer's hematoxylin.

Transgene expression is cytoplasmic (arrowheads), hematoxylin stains nuclei (arrows). **(D)**

PC 12 cells were differentiated by growth (at least 12 days) in serum-free medium

supplemented with 100 ng/ml of NGF and transfected with expression vector pJW17

(ICP10) (*solid triangles*) or pJHL15 (p139™) (*open triangles*). Non-transfected cells (*open squares*) served as control. Viability was determined by MTS assay (using the CellTiter 96 Aqueous One Solution Cell Proliferation Assay) and results were expressed as % relative to the number of cells at 0 hrs post NGF withdrawal $\pm$ SEM (*, p< 0.05 vs. control, +, p < 0.05 vs. pJHL15-transfected cells, by ANOVA). **(E)** Mouse euploid hippocampal neurons were plated on glass coverslips etched with a grid of 175 x 175 m squares, maintained for 2 days in MEM/B27 media, transfected with pJW17 (*solid triangles*) or pJHL15 (*open triangles*) and the medium was replaced with MEM free of serum or B27. Non-transfected cells were maintained in MEM (Eu-B27) (*open squares*) or MEM with B27 (Eu+B27) (*solid circles*). Viability was determined by counting live neurons and results are expressed as a % relative to the initial (0 hrs) number of viable cells $\pm$ SEM [*, p< 0.05 vs. control (Eu-B27), +, p< 0.05 vs. pJHL15-transfected cells, by ANOVA].

3.   ICP10 PK protects trisomy 16 (Ts16) hippocampal neurons from naturally occurring apoptotic death. The Ts16 mouse is considered to be a model of DS (trisomy 21) (Sawa, 1999). Ts16 is a genetic defect believed to confer increased vulnerability to neurodegeneration (Sawa, 1999). Cultured hippocampal neurons from the Ts16 mouse exhibit increased cell death relative to littermate euploid cells, even in the presence of adequate trophic support (Bambrick and Krueger, 1995). To examine whether ICP10 PK can promote survival in this system, primary hippocampal cultures from Ts16 mice [established as described (Bambrick and Krueger, 1995)] were transfected with pJW17 or pJHL15 at 2 days in culture and maintained in B27-supplemented medium for the duration of the experiment. Live cells were counted as described above and the results are expressed as % surviving cells $\pm$ SEM.

As previously described (Bambrick and Krueger, 1995; Sawa, 1999), non-transfected (Ts16) neurons evidenced an accelerated death rate (76.2 ± 2.3; 53.7 ± 2.8, and 41.4 ± 2.3 % survival at 24, 48 and 72 hrs, respectively) relative to euploid neurons maintained in B27-supplemented medium (Eu+B27) (93.8 ± 1.7, 88.7 ± 3.1, and 80.6 ± 2.7 % survival at 24, 48 and 72 hrs, respectively) ($p < 0.01$ by ANOVA) (Fig. 51 D). Similar cell death was seen for pJHL15-transfected Ts16 neurons (73.8 ± 3.0, 54.9 ± 3.2 and 43.8 ± 3.7, at 24, 48, and 72 hrs ($p > 0.05$ vs. non-transfected Ts16 neurons by ANOVA). By contrast, the survival of pJW17-transfected Ts16 neurons (93.5 ± 1.5; 83.9 ± 2, and 75.6 ± 2.2 % at 24, 48 and 72 hrs, respectively) was similar to that of euploid neurons maintained in B27-supplemented medium (Eu+B27) and significantly ($p < 0.001$ by ANOVA) higher than that of non-transfected Ts16 neurons. Degenerating neurons were present in non-transfected (Fig.51 A), pJHL15-transfected (Fig. 51 B) but not pJW17-transfected (Fig. 51 C) Ts16 cultures at 72 hrs post-transfection (5 days in culture). In all three systems (PC12, primary hippocampal neurons deprived of trophic support and Ts16 cultures) the transgenes were expressed equally well and the increased survival of pJW17-transfected relative to non-transfected cells was 20-35 %, consistent with the estimated transfection efficiency (25-35 %).

To examine whether the increased survival of pJW17-transfected Ts16 cultures is due to the ability of ICP10 PK to negatively regulate apoptosis, the cultures (non-transfected or transfected with pJW17 or pJHL15) were fixed with 4 % paraformaldehyde at 72 hrs post-transfection and examined for apoptosis by TUNEL. Apoptotic cells (characterized by a dark nuclear precipitate) and non-apoptotic cells (unstained or displaying a diffuse, light and uneven cytoplasmic staining) were counted in 5 randomly chosen microscopic fields (containing at least 250 cells). Results are expressed as % apoptotic cells $\pm$ SEM. The proportion of TUNEL-positive (apoptotic) cells in non-transfected (43 ± 3.4 %) and pJHL15-transfected (39 ± 1 %) cultures was significantly higher ($p < 0.01$, by ANOVA) than in pJW17-transfected (9.4 ± 1.1 %) or in euploid cultures maintained in MEM with B27 supplement (Eu+B27) (9.6 ± 2 %) (Fig 51 E). The lower proportion of dead cells determined by TUNEL as compared to cell counting, presumably reflects cell loss during fixation. Similar results were obtained in 3 independent experiments, suggesting that ICP10 PK blocks apoptotic death of Ts16 neurons.

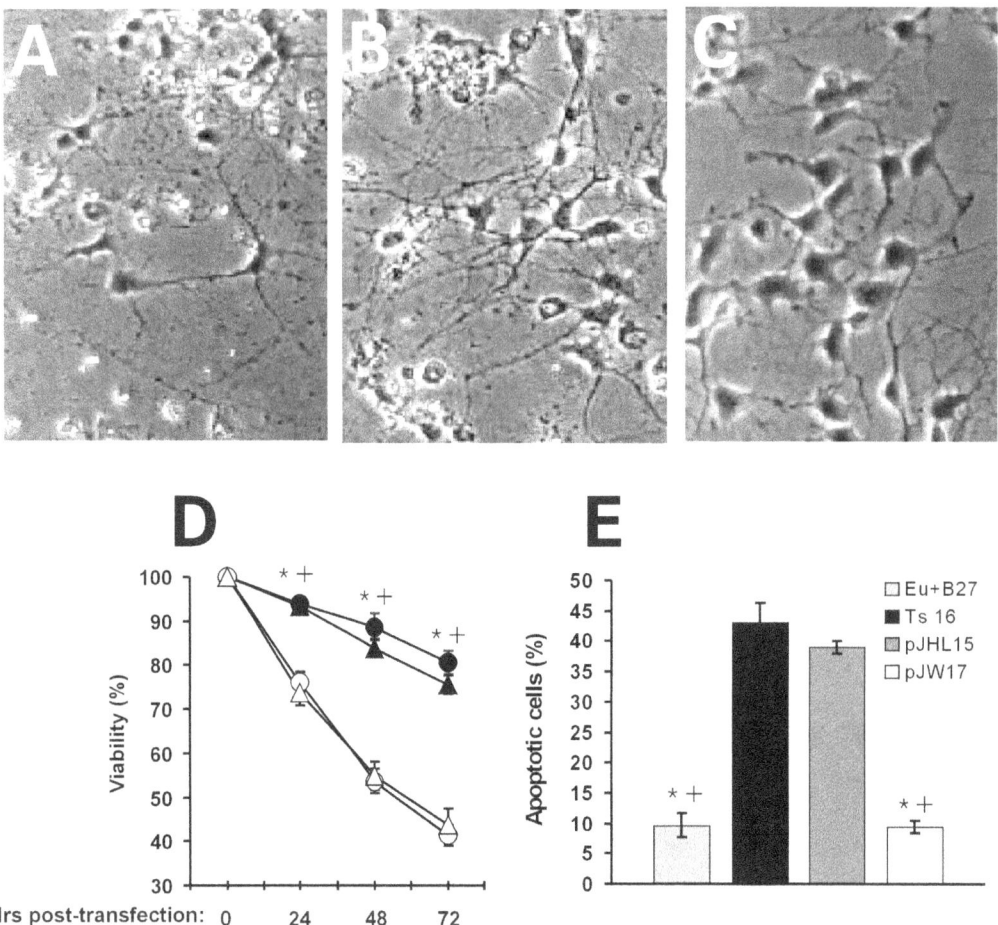

Fig. 51   ICP10 PK inhibits the naturally occurring apoptotic death of Ts16

hippocampal neurons. (A) Morphology of non-transfected Ts16 neurons at day 5 in

culture. (B) Morphology of pJHL15-transfected Ts16 neurons at day 5 in culture (3 days

post-transfection). (C) Morphology of pJW17-transfected Ts16 neurons at day 5 in culture

(3 days post-transfection). (D) Mouse Ts16 hippocampal neurons were grown on glass

coverslips and transfected with pJW17 (*solid triangles*), pJHL15 (*open triangles*) or non-transfected (*open circles*). Non-transfected euploid littermate hippocampal cells maintained in B27-supplemented medium (Eu+B27) (*solid circles*) were used as control. Viability was determined by counting live neurons and expressed as a % relative to the initial (t=0) number of viable cells $\pm$ SEM (*, p< 0.001 vs. control Ts16 non-transfected; (**E**) Ts16 mouse hippocampal neurons grown and transfected as in **D** and non-transfected euploid hippocampal cells maintained in B27-supplemented medium (Eu+B27) were assayed by TUNEL at 72 hrs post-transfection. Results are expressed as % apoptotic (TUNEL-positive) cells $\pm$ SEM (*, p< 0.01 vs. control Ts16 non-transfected; +, p< 0.01 vs. pJHL15-transfected cells, by ANOVA).

4.   Discussion-Specific Aim 4

Apoptosis is involved in the etiology of neurological disorders, such as acute brain injury (hypoxia-ischemia and trauma) and neurodegenerative disorders (AD, DS and ALS, and others).

Acute (or traumatic brain injury) is characterized by acute compromise of the blood-brain barrier, alterations in cerebral blood flow and metabolism, neurological motor dysfunction, and cognitive impairment (Conti *et al.*, 1998). Neuronal degeneration after experimental brain trauma can be detected as early as 10 min, may persist up to 1 year after injury (Smith *et al.,* 1997), and involves regionally distinct patterns of apoptotic death, involving the cortex and the hippocampus (Conti *et al.*, 1998). Hypoxia-ischemia, also has an apoptotic component (reviewed by Lipton, 1999).

AD affects approximately 4 million people in the US. Hallmarks of AD include the presence of senile plaques consisting of extracellular deposits of amyloid beta protein and selective loss of neurons due to apoptosis, especially in regions of the hippocampus that are associated with memory and learning leading to impaired cognitive and memory processes. Moreover, cholinergic neurons of the basal forebrain are also affected early in the disease due to the loss of target-derived neurotrophic factors (Honig and Rosenberg, 2000). DS is the most common clinical syndrome associated with mental handicap. It occurs in about 1 of 1,000 live births and accounts for about 15 % of the total mentally handicapped population. DS is associated with a full trisomy of chromosome 21, due to chromosomal non-disjunction during meiosis. There are common pathological mechanisms between AD and DS. Elderly persons with DS (as well as AD patients) have a lower number of neurons in the hippocampus, temporal cortex and entorhinal cortex, compared to similarly aged normal people in the general population, suggesting that the genetic defect in DS confers increased vulnerability to neurodegeneration. Moreover, DS patients develop AD by their fifth decade of life (Sawa, 1999).

ALS is a characterized by degeneration of motor neurons. About 3-4 % of cases are associated with mutations (about 50 are known) in the gene encoding the free radical scavenging enzyme Cu, Zn superoxide dismutase1 (SOD1) (Honig and Rosenberg, 2000).

The pro-apoptotic potential of mutant SOD1 is presumably due to an unstable conformation that leads to formation of toxic aggregates or to a gain in enzymatic function involving increased intracellular free radicals (Pasinelli et al., 1998).

The common denominator for these disorders is selective neuronal loss likely due to apoptosis, which is rare in the normal mature or aged brain. What exactly prompts apoptosis in chronic disorders is unclear, but several stimuli were postulated as potentially etiologic. They include: oxidative stress (that may increase with age), loss of neurotrophic support, genetic defects, accumulated burden of endogenous or exogenous factors, or excessive release of neurotransmitters known as excitotoxins.

To examine whether ICP10 PK can protect neurons from apoptotic stimuli specific for neurological disorders, several in vitro models were used, in which ICP10 or p139™ (a PK negative mutant of ICP10) were ectopically expressed by transient transfection.

As an in vitro model of ALS, we used mouse neuroblastoma cells that constitutively express wild type (wt) or mutant (G85R) SOD1 (Pasinelli et al., 1998) that were induced to differentiate to a neuronal phenotype (Pasinelli et al., 1998) and transfected with pJW17 and pJHL15 expression vectors (that respectively express ICP10 or p139™). Using two different assays (MTS and TUNEL) we showed that, ICP10 protected cells that express mutant SOD1 from oxidative stress (treatment with X/XO), by a PK-dependent mechanism. Significantly, expression of the mutant enzyme G85R did not cause apoptosis by itself, but conferred increased susceptibility to apoptosis induced by oxidative stress. This may reflect the in vivo situation, in which it is the accumulated burden of exogenous and endogenous factors that ultimately decides the fate of the cells expressing the mutant SOD1 enzymes, as previously suggested (Pasinelli *et al.*, 1998).

ICP10 PK also protected neuronally differentiated PC12 cells and primary cultures of hippocampal neurons from death due to trophic factor deprivation, an apoptotic stimulus known to occur in the development of many neurological disorders (including AD). As an *in vitro* model for DS and AD (Plioplys, 1991; Sawa. 1999) hippocampal cultures from the Ts16 mouse were used. Cultured hippocampal neurons from the Ts16 mouse exhibit increased cell death relative to littermate euploid cells, even in the presence of adequate trophic support (Bambrick and Krueger, 1995), suggesting that this genetic defect confers increased vulnerability to neurodegeneration (Sawa, 1999). ICP10 PK inhibited this naturally occurring apoptotic death due to the innate genetic defects, suggesting that it has a broad neuroprotective activity.

The salient feature of our findings is the observation that ICP10 PK blocks neuronal death in the experimental paradigms created by trophic factor deprivation or genetic defects (G85R SOD1 expression and Ts16). ICP10 has a transmembrane helical segment that serves to anchor it to the plasma membrane of HSV-2-infected and constitutively expressing cells (Luo and Aurelian, 1992). Its PK functions as an activated growth factor receptor, and both the auto- and trans-phosphorylating activities of a fusion protein consisting of ICP10 PK and the ligand binding extracellular domain of epidermal growth factor (EGF) receptor are ligand (EGF)-inducible (Smith *et al.*, 1996). Because ICP10 PK activates the Ras/Raf/MEK/ERK pathway in HSV-2-infected and constitutively expressing non-neuronal (Smith *et al.*, 1994; Smith *et al.*, 2000), and HSV-2-infected-neuronal cultures (present study), the most likely interpretation is that it blocks caspase-dependent apoptosis by activating the MEK/ERK survival pathway. The following findings are consistent with this interpretation: i) ICP10 PK protects hippocampal neurons from apoptosis induced by virus infection by inhibiting caspase-3 activation and activating the MEK/ERK pathway (Specific Aim 1 and Specific Aim 2), and ii) p139^{TM} does not activate the Ras/Raf/MEK/ERK pathway (Smith *et al.*, 1994) and does not rescue neurons from death due to apoptosis (Specific Aim 4). However, the exact contribution of the activated pathway to the neuroprotective activity of ICP10 PK in the paradigms described here (Specific Aim 4), the role of other survival pathways, and the identity of the survival genes involved in the ICP10PK-mediated protection, remain to be elucidated. Since survival was assessed *in toto*, future co-localization studies are also needed to determine whether individual cells that were not expressing ICP10 PK (due to the low transfection efficiency) are apoptotic (TUNEL-positive). Ongoing studies in our laboratory are designed to examine the mechanisms of cell death and survival in the paradigms described in this section.

In many neurological disorders (acute brain injury, neurodegenerative diseases), neurons die by apoptosis. The widespread nature of neuronal injury presents a considerable challenge to the development of therapeutic strategies. The development of strategies to prevent the death of CNS neurons is limited by the paucity of genes that block apoptosis in these cells and of non-invasive methods of delivery to the brain. By promoting the survival of hippocampal neurons through activation of intracellular survival pathways, ICP10 PK represents a novel therapeutic strategy for treatment of neurological disorders that involve apoptosis. However, delivery of therapeutic genes to the brain, in a safe, specific and non-invasive manner, represents a considerable challenge. Thus, the next series of experiments were designed to determine whether ICP10 PK can be safely delivered to the brain by an HSV-2-derived vector, using a non-invasive method of delivery.

Specific Aim 5. To establish a murine model of ICP10 PK gene delivery to the hippocampus, using intranasal infection with a growth-deficient HSV-2 mutant that expresses ICP10 PK.

The natural biology of HSV makes it attractive as a gene delivery vehicle for the central and peripheral nervous system, and experimental studies suggest it may be useful for the direct delivery of transgenes to other tissues as well. The vast majority of this research is however directed to cancer therapy (including brain tumors therapy) by using cytolytic HSV vectors carrying tumor death-inducing genes (Glorioso *et al.*, 1997). There are only a few reports of HSV vectors as carriers of therapeutic (including anti-apoptotic) genes in treatment of acute brain injury (e.g. ischemia) or chronic neurodegenerative disorders (e.g. PD) in animal models, but they involve invasive techniques such as stereotaxic injections into the affected areas of the brain (Antonawich *et al.*, 1999; Natsume *et al.*, 2001). However, all of the above mentioned efforts in developing gene therapy are using HSV-1 (but not HSV-2)-based vectors, and many of these vectors have been proven cytotoxic (Johnson *et al.*, 1992; Ho *et al.* 1995).

The use of mice inoculated intranasally with HSV (when the virus spreads to the CNS by way of olfactory or trigeminal nerves) was initially described by DeClercq and Luczak (1976) as a model for herpes encephalitis of humans (HSE), and has been used extensively to evaluate antiviral agents directed against this disease. Both HSV-1 and HSV-2 have been shown to induce encephalitis in mice, when inoculated intranasally with 100-1000 pfu, as early as 2 days p.i. (Meyding-Lamadé *et al.*, 1996; Meyding-Lamadé *et al.*, 1998; Jones *et al.*, 2000). Significantly (as mentioned before in this report), the brain structures damaged by HSE are part of an anatomical and functional neural unit, the limbic system. It severely involves hippocampal formation, including the subicular cortices and the amygdala, as well as the parahypocampal gyrus and the mesocortices (Damasio and Van Hoesen, 1985), largely the same areas affected in AD (Hyman *et al.*, 1984).

The complex multifactorial nature of neurological disorders that involve apoptosis requires a complex treatment approach which will have to target: i) the apoptotic cell death due to exogenous or endogenous stimuli, and ii) the decline in the cognitive processes. An effective treatment would have to be directed and made accessible to the brain, without causing neurological side effects, such as inflammation associated with meningitis and encephalitis.

The experiments presented in this section (Specific Aim 5) were designed to examine: i) the pattern of expression of ICP10 PK as delivered to the mouse brain following peripheral (non-invasive) administration using a growth compromised herpes virus vector (ICP10ΔRR) inoculated intranasally, and ii) the safety of the vector administration.

1. **Expression of ICP10 PK and LacZ in ICP10ΔRR-infected cells in culture.** The

construction of the HSV-2 mutant ICP10ΔRR was previously described (Peng *et al.*, 1996;

Smith *et al.*, 1998; Aurelian and Smith, 2000). It retains the N-terminal domain of the

ICP10 protein (that codes for the anti-apoptotic gene ICP10 PK) and it has the

ribonucleotide reductase (RR) domain replaced with the *LacZ* gene (Fig. 52).

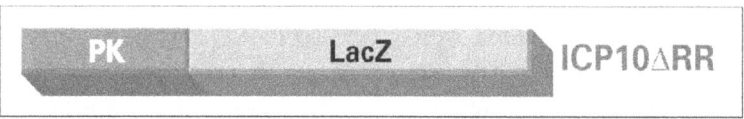

Fig. 52   **Schematic representation of ICP10ΔRR vector.**

Expression of the *LacZ* transgene in cultured cells, was examined by X-Gal

staining. Vero cells (grown in medium supplemented with 10 % FBS) were infected with

ICP10ΔRR (m.o.i. 10) or mock-infected with growth medium. At 24 hrs p.i. cells were

stained with X-Gal, as described in the Materials and Methods. ICP10ΔRR-infected cells

stained blue (Fig. 53 A), suggesting that the *LacZ* gene is expressed and its product (β-Gal

enzyme) is functional. It should be noted that in these cells and conditions (10 % FBS), the virus replicates (Aurelian and Smith, 2000), and the infected cells exhibit the specific

features associated with infection (e.g. rounding up) Mock-infected cells did not stain (Fig.

53 B).

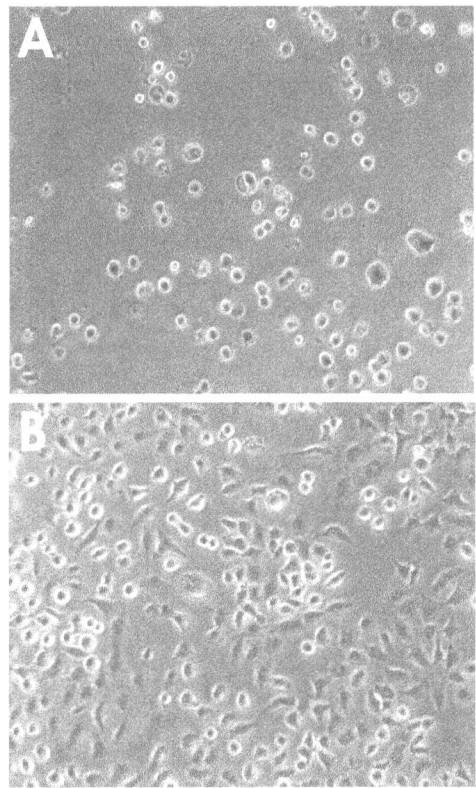

Fig. 53 **Expression of the *LacZ* transgene in cultured cells.** Vero cells were infected

with ICP10ΔRR (m.o.i 10) (A) or mock-infected (B). X-gal staining was done at 24 hrs

p.i.

ICP10ΔRR retains the ICP10 translation initiation site and expresses a 175 kDa (p175) fusion ICP10 PK-*LacZ* protein that retains its intrinsic PK activity (Aurelian and Smith, 2000). In the context of virus infection, ICP10 provides the PK activity that is necessary for IE gene expression and the RR activity required for viral DNA synthesis, thereby initiating the lytic cascade (Smith *et al.*, 1998; Aurelian and Smith, 2000). Consistent with this interpretation, mutants deleted in the RR domain of both HSV-1 and HSV-2, have been shown to be replication-defective in quiescent cells, and defective for reactivable virus (Jacobson *et al.*, 1989; Aurelian and Smith, 2000). Also, as described in this study, ICP10ΔRR is replication-defective in hippocampal cultures.

We showed earlier in this report that the p175 protein is expressed upon virion uncoating at 30 min p.i. To examine whether ICP10 PK is expressed at later times p.i. primary hippocampal cultures infected with ICP10ΔRR (m.o.i. 10) or mock-infected were stained by immunocytochemistry with ICP10 antibody that recognizes the p175 protein (Aurelian and Smith, 2000) at 24 hrs (data not shown) or 4 days p.i. (Fig. 54). Staining was seen at 24 hrs p.i. in ICP10ΔRR-infected cultures but not in mock-infected cultures at 24 hrs (data not shown) or 4 days p.i. (Fig. 54 B). As seen in Fig. 54 A, hippocampal cells infected with ICP10ΔRR maintained a healthy appearance and expressed the ICP10 PK at 4 days p.i., suggesting that ICP10ΔRR is not cytotoxic and it can be used for long-term gene expression in neuronal cultures.

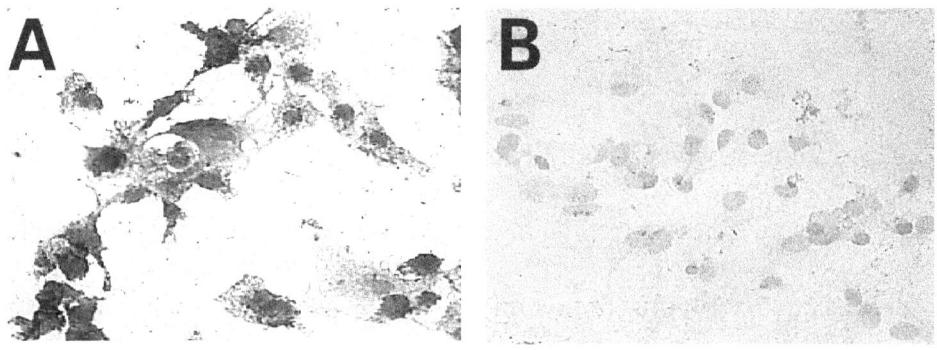

Fig. 54   ICP10 PK expression in hippocampal cultures. Primary cultures of rat

hippocampal neurons infected with ICP10ΔRR (m.o.i 10) (A), or mock-infected with

growth medium (B) were stained with antibody specific for ICP10, and counterstained

with hematoxylin at 4 days p.i.

2. ICP10 PK expression in the mouse brain following intranasal administration of ICP10ΔRR vector

To examine whether ICP10 PK is also expressed *in vivo*, six C57BL/6J mice (6-8 weeks of age) were anaesthetized with ether and infected intranasally by expelling 5 μl of virus suspension (5 x 10^6 pfu) from a micropipettor into each nostril. Three mice were mock-infected intranasally with sterile PBS (5 μl per nostril). Animals were examined daily concerning appearance (ruffled or normal coat), posture (hunched or normal), feeding habits, neurological signs (e.g. seizures, paresis, ataxia, circling), as previously described (Hudson *et al.*, 1991), and they were all normal. One mock- and one ICP10ΔRR-infected mouse were sacrificed at day 1 p.i., one mock- and two ICP10ΔRR-infected mice were sacrificed at 5 days p.i. and one mock- and three ICP10ΔRR-infected mice were sacrificed at 7 days p.i.. They were perfused with 2 % paraformaldehyde in 0.1 M Pipes buffer and their brains were dissected and infused with 20 % sucrose and frozen (-20 ^{0}C). Cryostat sections were cut in a coronal plane from the olfactory bulbs through the hippocampus, and stained with antibody specific for ICP10 by immunohistochemistry. Expression of ICP10 PK in the brain was consistent with a central spread of virus through the central olfactory pathways. Expression of ICP10 PK in the brain (as determined by immunohistochemistry with specific antibody and X-gal histochemistry) was consistent with a central spread of virus through the central olfactory pathways. Thus, staining of the olfactory bulb was seen as early as 1 day p.i. (not shown). At day 5 and 7 p.i., ICP10 PK expression was seen in the piriform cortex (Fig. 55 A, B), amygdala (Fig. 55 A), hippocampus (Fig. 55 D), neocortex (Fig. 55 D), striatum (Fig. 55 A, E), and entorhinal

cortex (not shown). Staining was not seen in the brains of mock-infected mice (Fig. 55 C,

E) or with rabbit normal serum (used as control). Evaluation of the brain with hematoxylin

and eosin (H & E) histology indicated that there was no neuronal degeneration or

infiltration of the brain or meninges by monocytes, macrophages or T lymphocytes (data

not shown). The inflammatory response was also evaluated by immunohistochemistry with

glial fibrillary acidic protein (GFAP)-specific antibody. Staining at 5 and 7 days p.i. did

not reveal any evidence of reactive astrocytosis in the path traveled by virus (data not

shown). These data suggest that: i) ICP10ΔRR delivers and expresses ICP10 PK in the

brain, upon intranasal infection of mice, and ii) ICP10ΔRR vector is safe and can be used

for long-term gene expression in the brain.

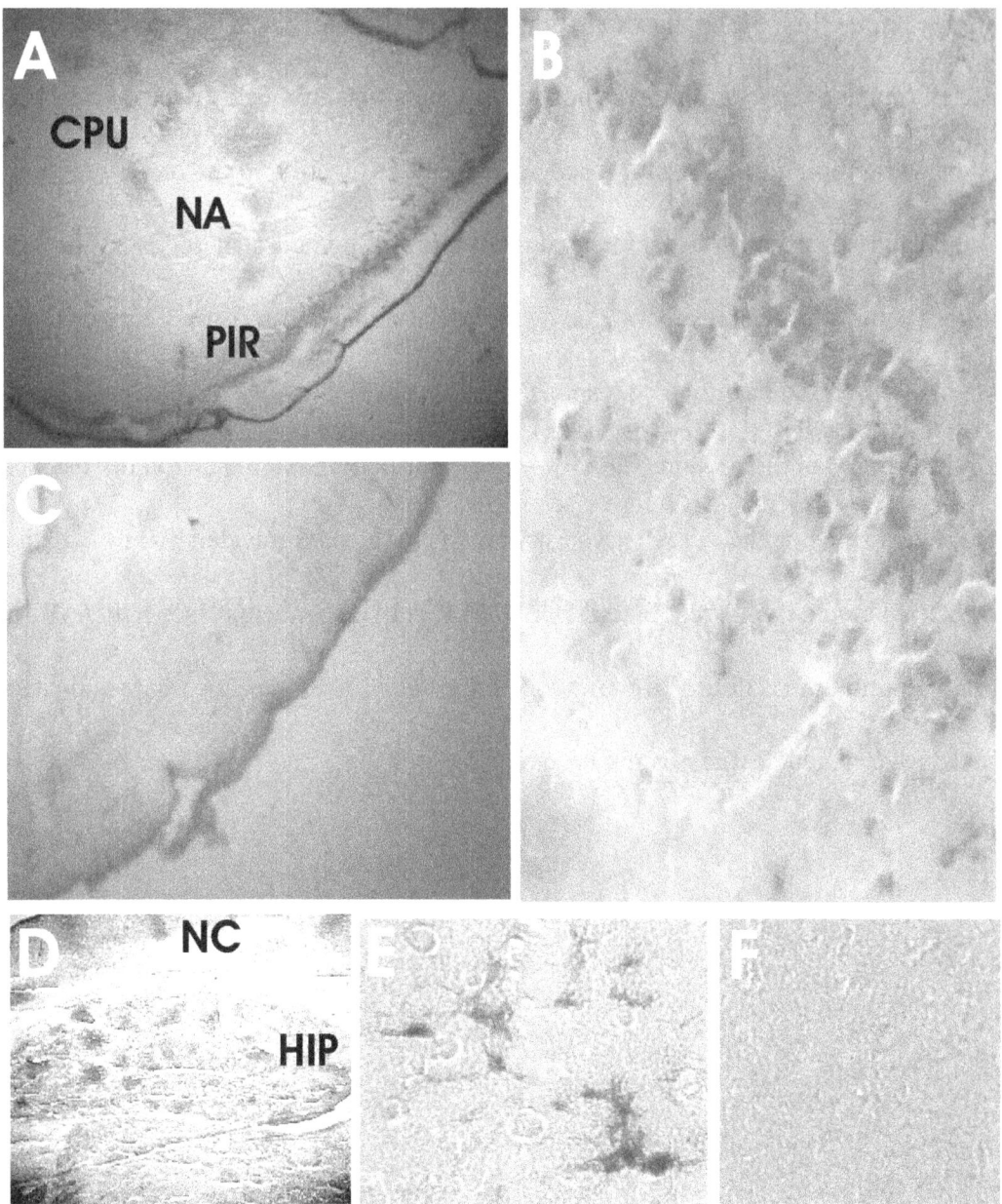

Fig. 55. ICP10 PK expression in the brains of mice infected intranasally with

ICP10ΔRR. Immunohistochemistry with ICP10-specific antibody at day 5 (A, C, D) and

day 7 p.i. (B, E, F) of brain sections from mice infected with ICP10ΔRR (A, B, D, E) or

PBS (control) (C, F). ICP10 PK expression was seen in the piriform cortex (PIR), amygdala (NA), striatum (CPU), hippocampus (HIP) and neocortex (NC) at 5 days p.i. Staining in the PIR and CPU is also seen at day 7 p.i. (B and E, respectively). Staining was not seen in brain sections from mock-infected mice (C, F).

3.   Discussion-Specific Aim 5

The features of ICP10 PK described in this study, as they relate to its abilities to inhibit virus-induced apoptosis and activate cellular survival pathways in hippocampal cultures, led us to the hypothesis that it may used in the treatment of neurological disorders that involve apoptosis. Thus, ICP10 PK has potential as a therapeutic gene, because: 1) it protects neurons from apoptosis, 2) it compensates for neurotrophic factor deprivation, and 3) it activates a neuronal survival pathway (MEK/ERK), which is involved in LTP- the molecular mechanism that underlies learning and memory. However, this approach also requires the means to deliver ICP10 PK to the brain in a safe and non-invasive manner. The experiments presented in this section (Specific Aim 5) were designed to examine whether ICP10 PK can be delivered and expressed in the brain following peripheral (non-invasive) administration using a growth compromised herpes virus vector (ICP10ΔRR) inoculated intranasally. Staining was seen in areas (olfactory bulb, piriform cortex, amygdala, entorhinal cortex, striatum and cortex) that suggest a central spread of the vector through the central olfactory pathways to the limbic system. Parallel experiments involving X-Gal immunohistochemistry, were done in the laboratory of Dr. Paul J. Yarowsky (University of Maryland, School of Medicine), and showed that the *LacZ* transgene is also expressed in the same areas. Significantly, these areas are also affected by neurodegeneration in human disease (Honig and Rosenberg, 2000). In some sections, staining had a "patchy" pattern that was previously described for HSV infection of the CNS (Wang *et al.*, 2001).

Before any virus can be used therapeutically, its capacity to cause disease must be abolished. In the case of HSV, its ability to cause encephalitis must be abrogated. In this sense, the present experiments showed that the ICP10ΔRR vector is safe, in that infected mice did not show any signs of neurological impairment or inflammation for up to 7 days p.i. This is significant since infection was done with doses much higher (x 5000-x 50.000) than those required to induce encephalitis in mice, as early as 2 days p.i. (Meyding-Lamadé *et al.*, 1996; Meyding-Lamadé *et al.*, 1998; Jones *et al.*, 2000).

Collectively, these data suggest that the ICP10ΔRR can be used as a vector for delivery of ICP10 PK to the brain, specifically to brain areas that are affected in neurological disorders that involve apoptosis.

Ongoing studies are designed in our laboratory and the laboratory of Dr. Paul J. Yarowsky (University of Maryland School of Medicine) to analyze the ICP10ΔRR-induced survival in neuronal cells in different apoptotic paradigms. These studies also involve characterization of ICP10ΔRR as a vector for the CNS. These studies aim to answer the questions that were not addressed in this report, such as duration of ICP10 PK expression, presence of an immune response after re-inoculation, vector dissemination to other organs, behavioral testing, and maintenance of synaptic transmission. Future experiments will also study the anti-apoptotic activity of ICP10ΔRR in various *in vivo* models of disease.

IV. GENERAL DISCUSSION

Viruses have evolved various strategies to inhibit apoptosis, including encoding of Bcl-2 homologs, inhibition of caspases or repression of p53 activity (Hardwick, 1997; Hardwick *et* al., 1998). Activation of a cellular survival pathway (PI3-K/Akt) by an anti-apoptotic viral gene (Hepatitis B virus- X protein) was also recently reported (Lee *et al.*, 2001). HSV-1 and HSV-2 have anti-apoptotic activity that is cell type specific and has been attributed to the HSV-1 and HSV-2 gene US3 and to the HSV-1 genes ICP34.5, US5, ICP27 and LAT, which function by a still poorly understood mechanism (Chou and Roizman, 1992; Aubert and Blaho, 1999; Hata *et al.*, 1999; Jerome *et al.*, 1999; Perng *et al.*, 2000). However, at this time (based on the experiments described in this report), ICP10 PK is the only HSV gene whose anti-apoptotic activity requires activation of a survival pathway (Raf/MEK/ERK). Moreover, no other anti-apoptotic viral genes that activate signaling pathways in hippocampal neurons, have been described.

Various anti-apoptotic viral genes have been found to be homologous to the cellular proteins involved in suppression of apoptosis (such as Bcl-2, IAP or FLIP) (Hardwick, 1997). ICP10 PK also has homology with a cellular protein from the heat shock family of proteins (Hsp) (Smith *et al.*, 1991) with anti-apoptotic activity (Gober *et* al., manuscript in preparation). The studies described in this report thus provide evidence for a novel mechanism of virus protection from apoptosis that involves coding for a gene

(ICP10 PK) homologous to an endogenous anti-apoptotic Hsp.

The present studies were based on previous observation that ICP10 PK activates the Ras/Raf/MEK/ERK mitogenic pathway in non-neuronal cells (Smith *et al.*, 1994; Smith *et al.*,2000). Since this pathway is involved in neuronal survival and function, and HSV-2 is neurotropic, it seemed of interest to determine whether ICP10 PK is involved in modulation of neuronal apoptosis. We found that, in neuronal cells, ICP10 PK has a broad anti-apoptotic activity. Its potential mechanism of action involves activation of the Raf/MEK/ERK survival pathway, leading to the expression of the anti-apoptotic protein Bag-1 (Fig. 56). Moreover, as Bag-1 has been shown to bind and activate the c-Raf kinase (Wang *et al.*, 1996), data suggest that, during infection, the newly synthesized Bag-1 may participate in activation of Raf/MEK/ERK pathway by a feedback mechanism. Since Bag-1 expression is also regulated by PI3-K/Akt pathway and ICP10 PK upregulates Akt activation, it is plausible that the mechanism of action of ICP10 PK may involve synergistic cooperation between Raf/MEK/ERK and PI3-K/Akt signaling pathways. The identity of the specific transcription factors involved in Bag-1 gene expression is not known. However, ICP10 PK also activates CREB, suggesting that this transcription factor may be a target of Raf/MEK/ERK survival pathway. In addition, ICP10 PK protected CREB from destruction by caspase cleavage. This is a significant finding in view of the CREB function without activation by phosphorylation. Thus, it has been reported that the non-phosphorylated CREB also has anti-apoptotic activity and can suppress AP-1 (c-Jun/c-fos heterodimers) transcriptional activity by competing for the AP-1 binding site on target genes (Masquilier and Sassone-Corsi, 1992). Moreover, ERK and CREB activation couple alterations in gene expression not only with survival but also with neuronal activity (Bailey *et al.*, 1996; Silva *et al.*, 1998), suggesting that ICP10 PK may also be involved in positive modulation of synaptic plasticity and cognitive functions.

Fig. 56   **Potential mechanism of action of ICP10 PK in hippocampal neurons.**

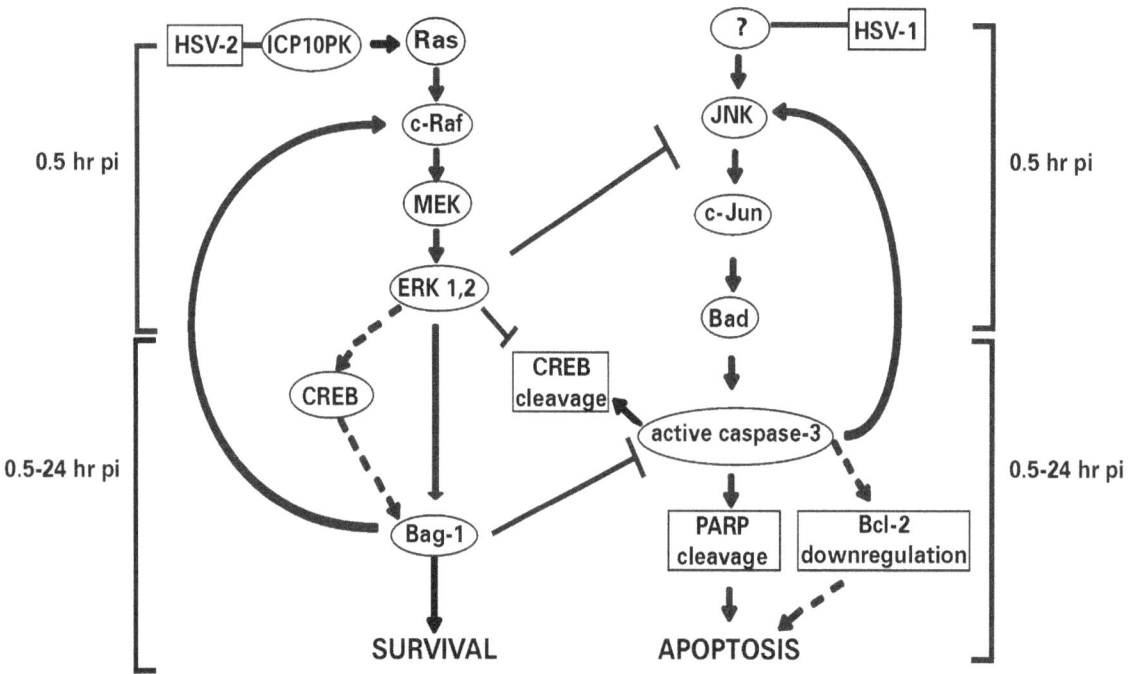

In addition to activating survival pathways, ICP10 PK is also involved in inhibition of pro-apoptotic factors, such as JNK pathway and its downstream targets. Thus, ICP10 PK inhibits the activation of JNK and its associated transcription factors associated with neuronal apoptosis (such as c-Jun, ATF-2, c-fos or p53). As discussed in Chapter I, sustained and/or elevated c-fos expression was associated with neuronal apoptosis and developmental failures both *in vitro* and *in vivo* (Ruther *et al.*, 1987; Smeyne *et al.*, 1993), ischemic brain injury was also associated with increases in c-fos, c-Jun and ATF-2 (Tong *et al.*, 1998; Walton *et al.*, 1998), and increased levels of p53 protein were associated with neuronal injury (Sakhi *et al.*, 1994), suggesting that these proteins can be used as markers of neuronal damage.

In contrast to ICP10 PK, our studies show that its HSV-1 homolog (ICP6 PK) does not have anti-apoptotic activity in hippocampal neurons. Moreover, the previously described anti-apoptotic genes encoded by HSV-1 (US3, LAT, ICP27, US5, US6, ICP27, ICP4) are not able to counteract virus-induced apoptosis. The potential mechanism of HSV-1-induced apoptosis is schematically described in Fig. 56. It involves activation of the JNK/c-Jun pro-apoptotic pathway, leading to expression of Bad. The identity of the pro-apoptotic viral gene(s) associated with this effect is not known. However, HSV-1-induced apoptosis is presumably associated with a change in the balance of pro- versus anti-apoptotic Bcl-2 family members in favor of the former, since in addition to induction of Bad expression, HSV-1 also induces downregulation of Bcl-2 protein levels. This process may lead to mitochondrial dysfunctions (Kaplan and Miller, 2000; Yuan and Yankner, 2000) and subsequent activation of caspases, the executioners of cell death. In addition, HSV-1 infection of hippocampal neurons failed to induce expression of the anti-apoptotic protein Bag-1, but ectopically expressed Bag-1 rescued the infected neurons from apoptosis, suggesting that Bag-1 plays an essential role in protection from virus-induced apoptosis.

What is the significance (if any) of HSV-1-induced apoptosis in CNS neurons ? In adults and older children, HSV-1 can cause HSE associated with severe destruction of temporal and frontal lobe structures which include the limbic mesocortices as well as amygdala and hippocampus (Whitley *et al*, 1982; Damasio and Van Hoesen, 1985). By contrast, HSV-2 infection of the CNS is most commonly restricted to a self-limiting, non-fatal meningitis (Bergstrom *et al*, 1990). The reason for this difference in CNS involvement between HSV-1 and HSV-2 is not known. Dissimilarities in routes of spread to the CNS have been proposed (Craig and Nahmias, 1973) but not confirmed (Kristensson *et al*, 1978; Corey *et al*, 1983). We hypothesize that a distinct effect on apoptosis by the two HSV types may be responsible for this distinct neurological outcome, as suggested by our results with primary cultures of cortical or hippocampal neurons that showed high levels of apoptosis in HSV-, but not HSV-2-infected cultures. Moreover, ectopically expressed ICP10 PK inhibited HSV-1-induced apoptosis, suggesting that its unique structural and functional features give HSV-2 the advantage of promoting neuronal survival, and presumably the less pathogenic effect associated with CNS infection.

HSE is associated with high mortality (30 %) even in the presence of adequate antiviral treatment and may still leave survivors with severe neurological impairment (Whitley and Lakeman, 1995; Skoldenberg, 1996). As such, improvements in HSE therapy are still important research goals. Our findings implicate apoptosis in HSE pathogenesis and raise the possibility of optimizing HSE therapy and possibly improving the neurological outcome of HSE patients by combining anti-viral with anti-apoptotic therapeutic agents.

In addition to virus infections of the CNS, apoptosis is also involved in the etiology of neurodegenerative disorders (AD, DS, ALS) and acute brain injury (trauma, ischemia/hypoxia). AD affects approximately 4 million people in the US. Hallmarks of AD include the presence of senile plaques consisting of extracellular deposits of amyloid beta protein and selective loss of neurons due to apoptosis, especially in regions of the hippocampus that are associated with memory and learning leading to impaired cognitive and memory processes. Moreover, cholinergic neurons of the basal forebrain are also affected early in the disease due to the loss of target-derived neurotrophic factors. DS is the most common clinical syndrome associated with mental handicap. It occurs in about 1 of 1,000 live births and accounts for about 15 % of the total mentally handicapped population. DS is associated with a full trisomy of chromosome 21, due to chromosomal non-disjunction during meiosis. There are common pathological mechanisms between AD and DS. Elderly persons with DS (as well as AD patients) have a lower number of neurons in the hippocampus, temporal cortex and entorhinal cortex, compared to similarly aged normal people in the general population, suggesting that the genetic defect in DS confers increased vulnerability to neurodegeneration. Moreover, DS patients develop AD by their fifth decade of life. ALS is a characterized by degeneration of motor neurons. About 3-4 % of cases are associated with mutations (about 50 are known) in the gene encoding the free radical scavenging enzyme Cu, Zn superoxide dismutase1 (SOD1) (Honig and Rosenberg, 2000). The pro-apoptotic potential of mutant SOD1 is presumably due to an unstable conformation that leads to formation of toxic aggregates or to a gain in enzymatic function involving increased intracellular free radicals (Pasinelli *et al.*, 1998). The common denominator for the neurodegenerative disorders is selective neuronal loss likely due to apoptosis which is rare in the normal mature or aged brain. What exactly prompts apoptosis in chronic disorders is unclear, but several stimuli were postulated as potentially etiologic. They include: oxidative stress that may increase with age, loss of neurotrophic support, accumulated burden of endogenous or exogenous factors, or excessive release of neurotransmitters known as excitotoxins.

The complex multifactorial nature of neurodegenerative disorders requires a complex treatment approach which will have to target: 1) the apoptotic cell death due to exogenous or endogenous stimuli, and 2) the decline in the cognitive processes. An effective treatment would have to be directed and made accessible to affected neuronal populations.

The novel concept described in this report refers to the use of the ICP10 PK as a therapeutic gene in the treatment of neurological disorders that involve apoptosis.

To examine whether ICP10 PK can protect neurons from apoptotic stimuli specific for neurological disorders, several *in vitro* models were used in which ICP10 or p139™ (a PK negative mutant of ICP10) were ectopically expressed by transient transfection.

As an *in vitro* model of ALS, we used mouse neuroblastoma cells that constitutively express wild type (wt) or mutant (G85R) SOD1 (Pasinelli *et al.*, 1998). In this system, ICP10 protected cells that express mutant SOD1 from oxidative stress, by a PK-dependent mechanism. ICP10 PK also protected neuronally differentiated PC12 cells and primary cultures of hippocampal neurons from death due to trophic factor deprivation, an apoptotic stimulus known to occur in the development of many neurological disorders (including AD). As an *in vitro* model for DS and AD (Plioplys, 1991; Sawa. 1999) hippocampal cultures from the Ts16 mouse were used. Cultured hippocampal neurons from the Ts16 mouse exhibit increased cell death relative to littermate euploid cells, even in the presence of adequate trophic support (Bambrick and Krueger, 1995), suggesting that this genetic defect confers increased vulnerability to neurodegeneration (Sawa, 1999). In this system, ICP10 PK inhibited the naturally occurring apoptotic death of Ts16 hippocampal cultures, suggesting that it has a broad neuroprotective activity that also covers the innate genetic defects associated with apoptosis. At this time, the mechanism by which ICP10 PK protects from apoptotic stimuli specific for neurological disorders is not known. However, it seems likely that it is mediated by ICP10 PK-induced activation of survival pathways (such as Ras/Raf/MEK/ERK) and specific transcription-dependent(or independent) events associated with this activation.

The use of ICP10 PK in therapy of neurological disorders would require a vector carrier that would ensure the expression of this therapeutic gene in the affected areas of the brain. Before any virus can be used therapeutically, its capacity to cause disease must be abolished. In the case of HSV, its ability to cause encephalitis must be abrogated. The HSV-2 mutant ICP10ΔRR was chosen as a vector because it is replication deficient, it has anti-apoptotic activity and it induces the expression of the anti-apoptotic Bag-1 protein. Moreover, ICP10ΔRR retains in its genome the ICP10 PK gene which activates the Ras/Raf/MEK/ERK which is required for long term potentiation (LTP)-the molecular mechanism of learning and memory (English and Sweatt, 1997). ICP10ΔRR also encodes for a transgene, *LacZ*, that can be used as a marker or can be replaced with another therapeutic gene of choice. To establish a murine model of ICP10 PK gene delivery to the brain, mice were inoculated intranasally with the ICP10ΔRR vector, and expression of ICP10 PK was determined by immunohistochemistry with specific antibody at various times p.i. ICP10 PK gene expression was seen in the hippocampus and other areas of the limbic system as late as 7 days p.i. However, the time pattern and duration of expression was not studied yet. The present experiments also showed that the ICP10ΔRR vector is safe, in that infected mice were normal and healthy and did not show any signs of neurological impairment or inflammation for up to 7 days p.i.

These results suggest that ICP10ΔRR can be used as a vector for the CNS. However, it remains to be seen whether ICP10ΔRR is also able to spare or improve neuronal function, and electrophysiological and behavioral studies will be needed to address these questions.

Currently, there are two main approaches for the treatment of AD.

1) The first approach, presumably a palliative one, is neurotransmitter replacement. It targets the cholinergic abnormalities occurring subsequent to hippocampal neurons degeneration but it does not address the cause of the cholinergic deficit which is hippocampal cell death. Studies of AD patients treated with cholinergic agonists (e.g. Bethanacol, Arecoline or Oxotremorine), cholinesterase inhibitors (e.g. tetrahydroaminoacridine or galanthamine) or cholinergic releasing agents (e.g. DuP 996) evidence only modest improvement of cognition.

2) The second is a neuroprotective approach, aimed at protecting neurons from further degeneration, thereby stopping cognitive deterioration. The use of pharmacological

approaches such as glutamatergic, chelating or anti-inflammatory agents as well as calcium channel blockers showed only a limited effect on the rate of cognitive decline. A more viable aim is the use of proteins capable of altering neuronal survival and function. Such a protein is the NGF which is expressed in the hippocampus, cortex and basal forebrain and it acts selectively on cholinergic neurons. NGF administration has been demonstrated to attenuate degenerative changes in cholinergic cells of primates caused by transection of the septohippocampal pathway. In humans, the intracerebroventricular administration of NGF induced improvement of verbal episodic memory, but no global cognitive improvement was observed. Moreover, it had severe side effects (Schwann cell hyperplasia, abundant sensory and sympathetic neurite sprouting around the brainstem and the spinal cord, weight loss, pain and confusion) due to the fact that the NGF mechanism of action is not restricted to the CNS neurons.

The proposed gene therapy using ICP10 PK would most likely combine the benefits of both approaches. Specifically, ICP10 PK protects hippocampal neurons from death, thereby inhibiting the neuronal loss. It also activates a neuronal survival pathway (MEK/ERK) which is also required for the molecular mechanisms of learning and memory. Significantly, ICP10 PK can be delivered to the affected areas by a non-invasive route (intranasally) thereby eliminating the need for intracerebroventricular administration.

Note

Provisional patent application regarding the use of ICP10 PK gene in the therapy of neurological disorders that involve apoptosis was filed by University of Maryland on July, 2001.

V.   REFERENCES

Alkondon, M. and Albuquerque, E.X.1993. Diversity of nicotinic acetylcholine receptors in rat hippocampal neurons. I. Pharmacological and functional evidence for distinct structural subtypes. *J.Pharmacol.Ex.Ther.* **265**: 1455-1473

Albuquerque, E.X., Alkondon, M., Pereira, E.F.R., Castro, N.G., Schrattenholz, A., Barbosa, C.T.F., Bonfante-Cabarcas, R., Aracava, Y., Eisenberg, H.M., Maelicke, A. 1997. Properties of neuronal nicotinic acetylcholine receptors: pharmacological characterization and modulation of synaptic function. *J. Pharmacol. Exp. Ther.* **280**: 1117-1136

Aloyz, R.S., Bamji, S.X., Pozniak, C.D., Toma, J.G., Atwal, J., Kaplan, D.R. and Miller, F.D. 1998. p53 is essential for developmental neuron death as regulated by TrkA and p75 neurotrophin receptors. *J. Cell. Biol.* **143**: 1691-1703

Amin, V., Cumming, D.V.E. and Latchman, D.S. 1996. Over-expression of heat shock protein 70 protects neuronal cells against both thermal and ischemic stress but with different efficiencies. *Neurosci. Lett.* **206**: 45-48

Anderson, A.J., Cummings, B.J., Cotman, C.W. 1994. Increased immunoreactivity for Jun- and Fos-related proteins in Alzheimer's disease: association with pathology. *Exp. Neurol.* **125**: 286-295

Anderson, A.J., Pike, C.J., and Cotman, C.W. 1995. Differential induction of immediate early gene proteins in cultured neurons by β-amyloid (Aβ): association of c-Jun with Aβ-induced apoptosis. *J. Neurochem.* **65**: 1487-1498

Andersson, T., Schultzberg, M., Schwarcz, R., Love, A., Wickman, C., and Kristensson, K. 1991. NMDA-receptor antagonist prevents measles virus-induced neurodegeneration. *Eur. J. Neurosci.* **3**: 66-71

Angel, P., Karin, M. 1991.The role of Jun, Fos and the AP-1 complex in cell-proliferation and transformation. *Biochim. Biophys. Acta* **1072**:129-157

Ankarcrona, M., Dypbukt, J.M., Bonfoco, E., Zhivotovsky, B., Orrenius, S., Lipton, S.A. and Nicoterra, P. 1995. Glutamate-induced neuronal death: a succession of necrosis or apoptosis depending on mitochondrial function. *Neuron* **15**: 961-973

Antonawich, F.J., Federoff, H.J., and Davis, J.N. 1999. Bcl-2 transduction, using a herpes simplex virus amplicon, protects hippocampal neurons from transient global ischemia. *Exper. Neurol.* **156**: 130-137

Arenas, E., Persson, H. 1994. Neurotrophin-3 prevents the death of adult central noradrenergic neurons in vivo. *Nature* **367**: 368-371

Arends, M.J., Morris, R.G., and Wyllie, A.H. 1990. Apoptosis: the role of the endonuclease. *Am.J.Pathol.* **136**: 593-608

Aronheim, A., Engelberg, D., Li, N., Al-Alawi, N., Schlessinger, J., and Karin, M. 1994. Membrane targeting of the nucleotide exchange factor is sufficient for activating the Ras signaling pathway. *Cell* **78** : 949-961

Asano, S., Honda, T., Goshima, F., Watanabe, D., Miyake, Y., Sugiura, Y., Nishiyama, Y. 1999. US3 protein kinase of herpes simplex virus type 2 plays a role in protecting corneal epithelial cells from apoptosis in infected mice. *J.Gen.Virol.* **80**: 51- 56

Atkins, C.M., Selcher, J.C., Petraitis, J.J., Trzaskos, J.M., Sweatt, J.D. 1998. The MAPK cascade is required for mammalian associative learning. *Nat. Neurosci.* **1**: 602-609

Aubert, M. and Blaho, J.A. 1999. The herpes simplex virus type 1 regulatory protein ICP27 is required for the prevention of apoptosis in infected human cells. *J. Virol.* **73**:2803-2813

Aubert, M., O'Toole, J., and Blaho, J.A .1999. Induction and prevention of apoptosis in human Hep-2 cells by herpes simplex virus type 1. *J. Virol.* **73**:10359-10370

Aubert, M., Rice, S.A. and Blaho, J.A. 2001. Accumulation of herpes simplex virus type 1 early and leaky-late proteins correlates with apoptosis prevention in infected human HEp-2 cells. *J. Virol.* **75**: 1013-1030

Auerbach, J.M., Segal, M. 1996. Muscarinic receptors mediating depression and long-term potentiation in rat hippocampus. *J. Physiol.* (London) **492**: 479-493

Aurelian, L., Terzano, P., Smith, C.C., Chung, T.D., Shamsuddin, A., Costa, S., and Orlandi, C. 1989.Amino-terminal epitope of herpes simplex virus type 2 ICP10 protein as a molecular diagnostic marker for cervical intraepithelial neoplasia. *Cancer Cells* **7**:187-191

Aurelian, L. 1998. Herpes simplex virus type 2 : unique biological properties include neoplastic potential mediated by the PK domain of the large subunit of ribonucleotide reductase. *Front.Biosci.* **3**: d237-d249

Aurelian, L., Kokuba, H., Smith, C.C. 1999. Vaccine potential of a herpes simplex virus type 2 mutant deleted in the PK domain of the large subunit of ribonucleotide reductase (ICP10). *Vaccine* **17**: 1951-1963

Aurelian, L. and Smith, C.C. 2000. Herpes simplex type 2 growth and latency reactivation by co-cultivation are inhibited with antisense oligonucleotides complementary to the translation initiation site of the large subunit of ribonucleotide reductase (RR1). *Antisensense & Nucleic Acid Drug Dev.* **10**: 77-85

Aurelian, L .2000. Herpes Simplex Viruses, in: *Clinical Virology Manual*, 3rd ed., Specter, S. Hodinka, R.L., Young, S.A. eds., ASM Press, Washington, DC, pp. 384-409

Aurelian, L., Smith, C.C., Winchurch, R., Kulka, M., Zaccaro, L., Gyotoku, T., Chrest, F.J., and Burnett, J.W.. 2001. A novel gene expressed in human keratinocytes with long term growth potential is required for cell growth. *J. Invest. Dermatol.* **116**: 286-295

Bailey, C.H., Bartsch, D. and Kandel, E.R. 1996. Toward a molecular definition of long term memory storage. *Proc. Natl. Acad. Sci. USA* **93**: 13445-13452

Balasz, R., Jorgensen, O.S. and Hack, N. 1988. N-methyl-D-aspartate promotes the survival of cerebellar granule cells. *Neuroscience* **27**: 437-451

Bambrick, L.L., Yarowsky, P.J., Krueger, B.K. 1995. Glutamate as a hippocampal neuron survival factor: an inherited defect in the trisomy 16 mouse. *Proc. Natl. Acad. Sci. USA* **92**: 9692-9696

Bambrick, L.L. and Krueger, B.K. 1999. Neuronal apoptosis in mouse trisomy 16: mediation by caspases. *J. Neurochem.* **72**: 1769-1772

Bar-Sagi, D., Feramisco, J.R. 1985. Microinjection of the ras oncogene protein into PC12 induces morphologic differentiation. *Cell* **42**: 81-84

Bardelli, A., Longati, P., Albero, D., Goruppi, S., Schneider, C., Ponzetto, C. and Comoglio, P.M.. 1996. HGF receptor associates with the anti-apoptotic protein BAG-1 and prevents cell death. *J. Biol. Chem.* **271**: 16850-16855

Behrens, A., M. Sibilis, E.F. Wagner. 1999. Amino-terminal phosphorylation of c-Jun regulates stress-induced apoptosis and cellular proliferation. *Nat Genet* **21**: 326-329

Beitner-Johnson, D., Guitart, X., Nestler, E.J.. 1992. Neurofilament proteins and the mesolimbic dopamine system: common regulation by chronic morphine and chronic cocaine in the rat ventral tegmental area. *J.Neurosci.* **12**: 2165-2176

Bergstrom, T., Svennerholm, B., Conradi, N., Horal, P. and Vahlne, A.1991. Discrimination of herpes simplex types 1 and 2 cerebral infections in a rat model. *Acta Neuropathol.* **82**: 395-401

Berman, D.E., Hazvi, S., Rosenblum, K., Seger, R., Dudai, Y. 1998. Specific and differential activation of mitogen-activated protein kinase cascades by unfamiliar taste in the insular cortex of the behaving rat. *J. Neurosci.* **18**: 10037-10044

Berns, E., Klijn, J.G.M., van Putten, W.L.J., de White, H.H., Look, M.P., Meijer-van Gelder, M.E., Willman, K., Portengen, H., Benraad, T.J., Foekens, J.A. 1998. p53 protein accumulation predicts poor response to tamoxifen therapy of patients with recurrent breast cancer. *J. Clin. Oncol.* **16**: 121-127

Bertollini, L., Ciotti, M.T., Cherubini, E. and Cattaneo, A. 1997. Neurotrophin-3 promotes the survival of oligodendrocyte precursors in embryonic hippocampal cultures under chemically defined conditions. *Brain Res.* **746**: 19-24

Blenis, J. 1993. Signal transduction via the MAP kinases: Proceed at your own RSK. *Proc. Natl. Acad. Sci. USA* **90**: 5889-5892

Bliss, T. and Lomo, T. 1973. Long-lasting potentiation of synaptic transmission in the dentate area of the anaesthetized rabbit following stimulation of the perforant path. *J. Physiol.* (London) **232**: 331-356

Block, T., Barney, S., Masonis, J., Maggioncalda, J., Valyi-Nagy, T., Fraser, N.W. 1994. Long term herpes simplex virus type 1 infection of nerve growth factor-treated PC12 cells. *J. Gen. Virol.* **75**: 2481-2487

Blum, S., Moore, A.N., Adams, F., Dash, P.K. 1999. A mitogen-activated protein kinase cascade in the CA1/CA2 subfield of the dorsal hippocampus is essential for long-term spatial memory. *J. Neurosci.* **19**: 3535-3544

Boise, L.H., Gonzales-Garcia, M., Postema, C.E., Ding, L., Lindsten, T., Turka, L.A., Mao, X., Nunez, G., and Thomson, C.B. 1993. Bcl-x, a bcl-2-related gene that functions as a dominant regulator of apoptotic cell death. *Cell* **74**: 597-608

Bolovan, C.A., Sawtell, N.M., and Thompson, R.L. 1994. ICP34.5 mutants of herpes simplex virus type 1 strain 17syn+ are attenuated for neurovirulence in mice and for replication in confluent primary mouse embryo cell cultures. *J. Virol.* **68**: 48-55

Bonni, A., Ginty, D.D., Dudek, H., Greenberg, M.E. 1995. Ser 133 phosphorylated CREB induces transcription via a cooperative mechanism that may confer specificity to neurotrophin signals. *Mol. Cell. Neurosci.* **6**: 168-183

Bonni, A., Brunet, A., West, A.E., Datta, S.R., Takasu, M.A., and Greenberg, M.E.1999. Cell survival promoted by the Ras-MAPK signaling pathway by transcription-dependent and independent mechanisms. *Science* **286**:1358-1362

Bossy-Wetzel, E., Bakiri, L. and Yaniv, M. 1997. Induction of apoptosis by the transcription factor c-Jun. *EMBO J.* **16**: 1695-1709

Bossy-Wetzel, E., Newmeyer, D.D., and Green, D.R.. 1998. Mitochondrial cytochrome c release in apoptosis occurs upstream of DEVD-specific caspase activation and independent of mitochondrial transmembrane depolarization. *EMBO J.* **17**: 37-49

Bothwell, M. 1991. Keeping track of neurotrophin receptors. *Cell* **65**: 915-918

Boulton, T.G., Nye, S.H., Robbins, D.J., Ip, N.Y., Radziejeweska, E., Morgenbesser, S.D., DePinho, R.A., Panayotatos, N., Cobb, M.H., and Yancopoulos, G.D. 1991. *Cell* **65**: 663-675

Brindle, P.K. and Montminy, M.R. 1992. CREB family of transcription activators. *Genes Dev.* **2**: 199-204

Brown, S., MacLean, A., Aitken, J. and Harland, J. 1994. ICP34.5 influences herpes simplex virus type 1 maturation and egress from infected cells in vitro. *J. Gen. Virol.* **75**: 3767-3686

Brunet, A., Bonni, A., Zigmond, M.J., Lin, M.Z., Juo, P., Hu, L.S., Anderson, M.J., Arden, K.C., Blenis, J., Greenberg, M.E. 1999. Akt promotes cell survival by phosphorylating and inhibiting a forkhead transcroption factor. *Cell* **96**: 857-868

Bump, N.J., Hackett, M., Hugunin, M., Seshagiri, S., Brady, K., Chen, P., Ferenz, C., Franklin, S., Ghayur, T., Li, P., et al. 1995. Inhibition of ICE family of proteases by baculovirus anti-apoptotic protein 35. *Science* **269**: 1885-1888

Bunnemann, B., A. Terron, V. Zantedeschi, P.E. Merlo, C. Chiamulera. 2000. Chronic nicotine treatment decreases neurofilament immunoreactivity in the rat ventral tegumental area. *Eur.J.Pharmacol.* **393**: 249-253

Caelles, C., Hemberg, A., Karin, M. 1994. p53-dependent apoptosis in the absence of transcriptional activation of p53-target genes. *Nature* **370**: 220-223

Charriaut-Marlangue, C., Aggoun-Zouaoui, D., Represa, A., Ben-Ari, Y. 1996. Apoptotic features of selective neuronal death in ischemia, epilepsy and gp120 toxicity. *Trends Neurosci.* **19**: 105-114

Chen, R.-H., Abate, C. and Blenis, J. 1993. Phosphorylation of the c-Fos transrepression domain by mitogen-activated protein kinase and 90-kDa ribosomal S6 kinase. *Proc. Natl. Acad. Sci. USA.* **90**: 10952-10956

Cheng, H.H.-Y., Nicholas, J., Bellows, D.S., Hayward, G.S., Guo, H.-G., Reitz, M.S. and Hardwick, J.M. 1997. A Bcl-2 homologue encoded by Kaposi's sarcoma-associated virus, human herpesvirus 8, inhibits apoptosis but does not heterodimerize with Bax or Bak. *Proc.Natl.Acad.Sci.USA.* **94**: 690-694

Chittenden, T. 1998. Mammalian Bcl-2 family genes, in *Apoptosis genes*, J.W. Wilson, C. Booth, C.S. Potten, eds., Kluwer Acad. Publishers, pp. 37-83

Choi, D. 1988. Glutamate neurotoxicity and diseases of the nervous system. *Neuron* **1**: 623-634

Choi, D.W. 1995. Calcium: still center-stage in hypoxic-ischemic neuronal death. *Trends Neurosci.* **18**: 58-60

Chou, J., Kern, E., Whitley, R. and Roizman, B.1990. Mapping of herpes simplex virus neurovirulence to γ1 34.5, a gene nonessential for growth in culture. *Science* **250**:

1262-1265

Chou, J. and Roizman, B. 1992. The gamma$_1$34.5 gene of herpes simplex virus 1 precludes neuroblastoma cells from triggering total shut off of protein synthesis characteristic of programmed cell death in neuronal cells. *Proc. Natl. Acad. Sci. USA* **89**: 3266-3270

Chou, J. and Roizman, B. 1994. Herpes simplex virus 1 γ_1 34.5 gene function, which blocks the host response to infection maps in the homologous domain of the genes expressed during growth arrest and DNA damage. *Proc.Natl.Acad.Sci.USA* **91**: 5247-5251

Chung, T.D., Wymer, J.P., Smith, C.C., Kulka, M., and Aurelian, L.1989. Protein kinase activity associated with the large subunit of herpes simplex virus type 2 ribonucleotide reductase (ICP10). *J.Virol.* **63**: 3389-3398

Claret, F.X., Hibi, M., Dhut, S., Toda, T., and Karin, M. 1996. A new group of conserved coactivators that increase the specificity of AP-1 transcription factors. *Nature* **383**: 453-457

Clarke, P.B., Schwartz, R.D., Paul, S.M., Pert, C.B. and Pert, A. 1985. Nicotinic binding in rat brain: autoradiographic comparison of [3H] acetylcholine, [3H] nicotine, and [125I]-alpha-bungarotoxin. *J. Neurosci.* **5**: 1307-1315

Clem, R.J., Hardwick, J.M. and Miller, L.K. 1996. Anti-apoptotic genes of baculoviruses. *Cell. Death Diff.* **3**: 9-16

Clem, J., Cheng, E.H.-Y., Karp, C.L., Kirsch, D.G, Ueno, K., Takahashi, A., Kastan, M.B., Griffin, D.E., Earnshaw, W.C., Veliuona, M.A. and Hardwick, J.M. 1998.

Modulation of cell death by bcl-xL through caspase interaction. *Proc.Natl.Acad.Sci. USA* **95**: 554-559

Cleveland, D.W. 1999. From charcot to SOD1: mechanisms of selective motor neuron death in ALS. *Neuron* **24**: 515-520

Clevenger, C.V., Thickman, K., Ngo, W., Chang, W.-P., Takayama, S., and Reed, J.C. 1997. Role of Bag-1 in the survival and proliferation of the cytokine-dependent lymphocyte lines, Ba/F3 and Nb2. *Mol. Endocrinol.* **11**: 608-618

Cohen, G.M., Sun, X.-M., Snowden, R.T., Dinsdale, D. and Skilleter, D.N. 1992. Key morphological features of apoptosis may occur in the absence of internucleosomal DNA fragmentation. *Biochem. J.* **286**: 331-334

Cohen, A., Bray, G., Aguayo, A. 1994. Neurotrophin 4/5 (NT 4/5) increases adult rat retinal ganglion cell survival and neurite outgrowth in vitro. *J. Neurobiol.* **25**: 953:959

Collingridge, G.L., Keal, S.J., Mclennan, H. 1983. Excitatory amino acid in synaptic transmission in the schffer collateral-commissural pathway in the rat hippocampus. *J. Physiol.* (London) **334**: 33-46

Collingridge, G.L. and Bliss, T.V.P. 1987. NMDA receptors-their role in long-term potentiation. *Trends Neurosci.* **10**: 288-293

Conner, J., Cooper, J., Furlong, J. and Clements, J.B. 1992. An autophosphorylating but not transphosphorylating activity is associated with the unique N-terminus of the herpes simplex virus type 1 ribonucleotide reductase large subunit. *J.Virol.* **66**: 7511-7516

Conti, A.C., Raghupathi, R., Trojanowski, J.Q., McIntosh, T.K. 1998. Experimental brain injury induces regionally distinct apoptosis during the acute and delayed post-traumatic period. *J. Neurosci.* **18**: 5663-5672

Cookson, M.R., Ince, P.G., Usher, P.A., Shaw, P.J. 1999. Poly (ADP-ribose) polymerase is found in both the nucleus and cytoplasm of human CNS neurons. *Brain Res* **834**: 182-185

Cooper, J., Conner, J., and Clements, J.B. 1995. Characterization of the novel protein kinase activity present in the R1 subunit of herpes simplex virus ribonucleotide reductase. *J. Virol.* **69**: 4979-4985

Cordone, M.H., Salvesen, G.S., Widmann, C., Johnson, G. and Frisch, S.M. 1997. The regulation of anoikis: MEKK-1 activation requires cleavage by caspases. *Cell* **90**: 315-323

Cotman, C.W., Monaghan, D.T. and Ganong, A.H. 1988. Excitatory amino acid receptors and neurotransmission: NMDA receptors and Hebb-type synaptic plasticity. *Annu. Rev. Neurosci.* **11**: 61-80

Court J, Martin-Ruiz C, Piggott M, Spurden D, Griffiths M, Perry E. 2001. Nicotinic receptor abnormalities in Alzheimer's disease. *Biol. Psychiatry* **49**: 175-184

Cowley, S., Patterson, H., Kemp, P., Marshall, C. 1994. Activation of MAP kinase is necessary and sufficient for PC12 differentiation and for transformation of NIH 3T3 cells. *Cell* **77**: 841-852

Cuenda, A., Rouse, J., Doza, Y.N., Meier, R., Cohen, P., Gallagher, T.F., Young, P.R., and Lee, J.C. 1995. SB 203580 is a specific inhibitor of a MAP kinase homologue which is stimulated by cellular stresses and interleukin-1. *FEBS Lett* **364**: 229-233

Culmsee, C., Zhu, X., Yu, Q.-S., Chan, S.L., Camandola, S., Guo, Z., Greig, N.H. and Mattson, M.P. 2001. A synthetic inhibitor of p53 protects neurons against death induced by ischemic and excitotoxic insults, and amyloid β-peptide. *J. Neurochem.* **77**: 220-228

Dahl, J., Jurczak, A., Cheng, L.A., Baker, D.C., Benjamin, T.L. 1998. Evidence of a role for phosphatidylinositol 3-kinase activation in the blocking of apoptosis by polyomavirus middle T antigen. *J. Virol.* **72**: 3221-3226

Damasio, A.R., and Van Hoesen, G.W. 1985. The limbic system and the localization of herpes simplex encephalitis. *J. Neurol. Neurosurg. Psych.* **48**: 297-301

Danaher, R.J., Jacob, R.J., Chorak, M.D., Freeman, C.S., and Miller, C.S. 1999. Heat stress activates production of herpes simplex virus type 1 from quiescently infected neurally differentiated PC12 cells. *J. NeuroVirol.* **5**: 374-383

Datta, S.R., Dudek, H., Tao, X., Masters, S., Fu, H., Gotoh, Y. and Greenberg, M.E. 1997. Akt phosphorylation of BAD couples survival signals to the cell-intrinsic death machinery. *Cell* **91**: 231-241

Datta, S. R., Brunet, A. and Greenberg, M. E. 1999. Cellular survival: a play in three Akts. *Genes Dev.* **13**: 2905-2927

Daum, G., Eisenmann-Tappe, I., Fries, H.W., Troppmair, J. and Rapp, U.R. 1994. The ins and outs of Raf kinases. *Trends Biochem. Sci.* **19**: 474-480

De Clercq, E., Luczak, M. 1976. Intranasal challenge of mice with herpes simplex virus: an experimental model for evaluation of the efficacy of antiviral drugs. *J. Infect. Dis.* **133** (suppl): A226-A236

Deng, X., Ruvolo, P., Carr, B., and May, W.S. Jr. 2000. Survival function of ERK1/2 as IL-3-activated, staurosporine-resistant Bcl2 kinases. *Proc. Natl. Acad. Sci.* **97**: 1578-1583

de Rijk, M.C., Breteler, M.M., Graveland, G.A., Ott. A., Grobbee, D.E., van der Meche, F.G., Hofman, A. 1995. Prevalence of Pakinson's disease in the elderly: the Rotterdam Study. *Neurology* **45**: 2143-2146

Dent, P., Reardon, D.B., Morrison, D.K., and Sturgill, T.W. 1995. Regulation of Raf-1 and Raf-1 mutants by Ras-dependent and Ras-independent mechanisms in vitro. *Mol. Cell. Biol.* **15**: 4125-4135

Dever, T.E. 1999. Translation initiation: adept at adapting. *Trends Biochem. Sci.* **24**: 398-403

Deveraux, Q.L., Takahashi, R., Salvesen, G.S., Reed, J.C. 1997. X-linked IAP is a direct inhibitor of cell-death proteases. *Nature* **388**: 300-304

Dieken, E.S. and Miesfeld, R.L. 1992.Transcriptional transactivation functions localized to the glucocorticoid receptor N terminus are necessary for steroid induction of lymphocyte apoptosis. *Mol. Cell. Biol.* **12**: 589-597

Dineley, K.T, Westerman, M., Bui, D., Bell, K., Hsiao Ashe, K., and Sweatt, J.D. 2001. β-amyloid activates the mitogen-activated protein kinase cascade via hippocampal α7 nicotinic acetylcholine receptors: *in vitro* and *in vivo* mechanisms related to Alzheimer's disease. *J. Neurosci.* **21**: 4125-4133

Djebaili, M., Rondoin, G., Baille, V. and Bockaert, J. 2000. p53 and Bax implication in NMDA induced-apoptosis in mouse hippocampus. *NeuroReport* **11**: 2973-2976

Dobrowsky, R., Werner, M., Castellino, A., Chao, M., Hannun, Y. 1994.Activation of the sphingomyelin cycle through the low affinity neurotrophin receptor. *Science* **265**: 1596-1599

Dorstyn, L., Kinoshita, M. and Kumar, S. 1998. Caspases in cell death, in *Apoptosis: mechanisms and role in disease*, S. Kumar ed., Berlin- New York; Springer, pp 1-23

Downward, J. 1998. Ras signaling and apoptosis. *Curr. Opinion Gen. Dev.* **8**: 49-54

Du, K. and Montminy, M. 1998. CREB is a regulatory target for the protein kinase Akt/PKB. *J. Biol. Chem.* **273**: 32377-32379

Durten-van Oorchot, A.A.A.M., den Hollander, A., Takayama, S., Reed, J.C., van der Eb, A.J., and Noteborn, M.H.M. 1997. Bag-1 inhibits p53-induced but not apoptin-induced apoptosis. *Apoptosis* **2**: 395-402

English, J.D. and Sweatt, J.D. 1997. A requirement for the mitogen-activated protein kinase cascade in hippocampal long term potentiation. *J. Biol. Chem.* **272**: 19103-19106

Erhardt, P., Schremser, E.J., Cooper, G.M. 1999. B-Raf inhibits programmed cell death downstream of cytochrome c release from mitochondria by activating the MEK/ERK pathway. 1999. *Mol. Cell. Biol.* **19**: 5308-5315

Esiri, M.M. 1982. Herpes simplex encephalitis. *J.Neurol.Sci.* **54**: 209-226

Estus, S., Zaks, W.J., Freeman, R.S., Gruda, M., Bravo, R., and Johnson, E.M. Jr. 1994. Altered gene expression in neurons during programmed cell death: identification of c-jun as necessary for neuronal apoptosis. *J. Cell. Biol.* **127**: 1717-1727

Estus, S., Tucker, H.M., van Rooyen, C., Wright, S., Brigham, E.F., Wogulis, M. and Rydel, R.E. 1997. Aggregated amyloid-β protein induces cortical neuronal apoptosis and concomitant "apoptotic" pattern of gene induction. *J. Neurosci.* **17**: 7736-7745

Favata, M.F., Horiuchi, K.Y., Manos, E.J., Daulerio, A.J., Stradley, D.A., Feeser, W.S., Van Dyk, D.E., Pitts, W.J., Earl, R.A., Hobbs, F., Copeland, R.A., Magolda, R.L., Scherle, P.A., and Trzaskos, J.M .1998. Identification of a novel inhibitor of mitogen-activated protein kinase kinase. *J. Biol. Chem.* **273**: 18623-18632

Fereira, A. and Caceres, A. 1992. Expression of the class III beta-tubulin isotype in developing neurons in culture. *J.Neurosci.Res.* **32**: 516-529

Fernandez-Alnemri, T., Litwack, G., and Alnemri. E. 1994. CPP32, a novel human apoptotic protein with homology to Caenorhabditis elegans cell death protein Ced-3 and mammalian interleukin-1 beta converting enzyme. *J. Biol. Chem.* **269**: 30761-30764

Finkbeiner, S., Tavazoie, S.F., Maloratsky, A., Jacobs, K.M., Harris, K.M., and Greenberg, M.E. 1997. CREB: a major mediator of neuronal neurotrophin responses. *Neuron* **19**: 1031-1047

Fiore, R.S., Bayer, V.E., Pelech, S.L., Posada, J., Cooper, J.A. and Baraban, J.M. 1993. *Neuroscience* **55**: 463-472

Fink, D.J. and Glorioso, J.C. 1997. Engineering herpes simplex virus vectors for gene transfer to neurons. *Nature Med.* **3**: 357-359

François, F., and Grimes, M.L. 1999. Phosphorylation-dependent Akt cleavage in neural cell in vitro reconstitution of apoptosis. *J. Neurochem.* **73**: 1773-1776

François, F., Godinho, M.J., Grimes, M.L. 2000. CREB is cleaved by caspases during neural cell apoptosis. *FEBS Lett* **486**: 281-284

Frazier, D.P., Cox, D., Godshalk, E.M. and Schaffer, P.A. 1996. The herpes simplex virus type 1 latency-associated transcript promoter is activated through Ras and Raf by nerge growth factor and sodium butyrate in PC12 cells. *J. Virol.* **70**: 7424-7432

Freed, E., Symons, M., MacDonald, S.G., McCormick, F., and Ruggieri, R. 1994. Binding of 14-3-3 proteins to the protein kinase Raf and effects on its activation. *Science* **265**: 1713-1716

Fuchs, S.Y., Adler, V., Pincus, M.R. and Ronai, Z. 1998. MEKK/JNK signaling stabilizes and activates p53. *Proc. Natl. Acad. Sci. USA* **95**: 10541-10546

Fukasawa, K., Rulong, S., Resau, J., Pinto da Silva P., Woude, G.F. 1995. Overexpression of mos oncogene product in Swiss 3T3 cells induces apoptosis preferentially during S-phase. *Oncogene* **10**: 1-8

Fukasawa, K., Vande Woude, G.F. 1997. Synergy between the mos-mitogen-activated protein kinase pathway and loss of p53 function in transformation and chromosome instability. *Mol. Cell. Biol.* **17**: 506-518

Galvan, V., and Roizman, B. 1998. Herpes simplex virus 1 induces and blocks apoptosis at multiple steps during infection and protects cells from exogenous inducers in a cell-type-dependent manner. *Proc. Natl. Acad. Sci. USA* **95**: 3931-3936

Galvan, V., Brandimarti, R. and Roizman, B. 1999. Herpes simplex virus 1 blocks caspase 3 - independent and caspase - dependent pathways to cell death. *J.Virol.* **73** : 3219 - 3226

Galvan, B., Brandimarti, R., Munger, J., Roizman, B. 2000. Bcl-2 blocks a caspase - dependent pathway of apoptosis activated by herpes simplex virus infection in HEp-2 cells. *J.Virol.* **74**: 1931 -1938

Gardoni, F., Schrama, L.H., Kamal, A., Gispen, W.H., Cattabeni, F., DiLuca, M. 2001. Hippocampal synaptic plasticity involves competition between Ca $^{2+}$/calmodulin-dependent protein kinase II and postsynaptic density 95 for binding to the NR2A subunit of the NMDA receptor. *J. Neurosci.* **21**: 1501-1509

Gavrieli, Y., Sherman, Y., and Ben-Sasson, S.A. 1992. Identification of programmed cell death via specific labeling of nuclear DNA fragmentation. *J.Cell.Biol.* **119**: 493-501

Gillardon, F., Spranger, M., Tiesler, C. and Hossman, K.-A. 1999. Expression of cell death-associated phosph-c-Jun and p53-activated gene 608 in hippocampal CA1 neurons following global ischemia. *Mol. Brain Res.* **73**: 138-143

Ginty, D.D., Bonni, A., Greenberg, M.E. 1994. Nerve growth factor activates a Ras-dependent protein kinase that stimulates c-fos transcription via phosphorylation of CREB. *Cell* **77**: 713-725

Giulian, D., Vaca, K., and Noonan, C.A. 1990. Secretion of neurotoxins by mononuclear phagocytes infected with HIV-1. *Science* **250**: 1593-1596

Giovannini, M.G., Blitzer, R.D., Wong, T., Asoma, K., Tsokas, P., Morrison, J.H., Iyengar, R., Landau, E.M. 2001. Mitogen-activated protein kinase regulates early phosphorylation and delayed expression of Ca^{2+}/calmodulin-dependent protein kinase II in long-term potentiation. *J. Neurosci.* **21**: 7053-7062

Glorioso, J.C., Goins, W.F., Schmidt, M.C., Krisky, D.M., Marconi, P.C., Cavalcoli, J.D., Ramakrishnan, R., Poliani, P.L., Fink, D.J. 1997. Engineering herpes simplex virus vectors for human gene therapy. *Adv. Pharmacol.* **40**: 103-136

Gold, R., Schmied, M., Giegerich, G., Beitschopf, H., Hartung, H.P., Toyka, K.V. and Lassman, H. 1994. Differentiation between cellular apoptosis and necrosis by the combined use of in situ tailing and nick translation techniques. *Lab. Invest.* **71**: 219-225

Goldstein, D.J. and Weller, S.K. 1987. Herpes Simplex Virus Type 1-induced ribonucleotide reductase activity is dispensable for virus growth and DNA synthesis: isolation and characterization of an ICP6 lacZ insertion mutant. *J. Virol.* **62**: 196-205

Goldstein, D.J. and Weller, S.K.. 1988. Factor(s) present in Herpes Simplex Virus Type 1-infected cells can compensate for the loss of the large subunit of the viral ribonucleotide reductase: characterization of an ICP6 deletion mutant. *Virology* **166**: 41-51

Goltsev, Y.V., Kovalenko, A.V., Arnold, E., Varfolomeev, E.E., Brodianskii, V.M., Wallach, D. 1997. CASH, a novel caspase homologue with death effector domains. *J. Biol. Chem.* **272**: 19641-19645

Goslin, K. and Banker, G. 1998. Rat hippocampal neurons in low-density culture, in *Culturing Nerve Cells*, G. Banker and K. Goslin, eds., pp 251-281

Grandgirard, D., Studer, E., Monney, L., Belser, T., Fellay, I., Borner, C., and Michel, M.R.. 1998. Alphaviruses induce apoptosis in Bcl-2-overexpressing cells: evidence for a caspase-mediated, proteolytic inactivation of Bcl-2. *EMBO J.* **17**: 1268-1278

Greene, L.A. 1978. Nerve growth factor prevents the death and stimulates neuronal differentiation of clonal PC12 pheochromocytoma cells in serum-free medium. *J.Cell.Biol.* **78**: 747-755

Greene, L.A. and Tischler, A.S. 1982. PC12 pheocromocytoma cells on neurobiological research. *Adv.Cell.Neurobiol.* **3**: 373-414

Greene, D.A., Stevens, M.J., Obrosova, I. and Feldman, E.L. 1999. Glucose-induced oxidative stress and programmed cell death in diabetic neuropathy. *Eur. J. Pharmacol.* **375**: 217-223

Grewal, S.S., Horgan, A.M., York, R.D., Withers, G.S., Banker, G.A., Stork, P.J. 2000. Neuronal calcium activates a Rap1 and B-Raf signaling pathway via the cyclic adenosine monophosphate-dependent protein kinase. *J. Biol. Chem.* **275**: 3722-3728

Guo, Q., Sopher, B.L., Furukawa, K., Pham, D.G., Robinson, N., Martin, G.M. and Mattson, M.P.1997. Alzheimer's presenilin mutation sensitizes neural cells to apoptosis induced by trophic factor withdrawal and amyloid beta-peptide: involvement of calcium and oxyradicals. *J. Neurosci.* **17**: 4212-4222

Gupta, S., Campbell, D., Derijard, B., Davis, R.J. 1995. Transcription factor ATF2 regulation by the JNK signal transduction pathway. *Science* **267**: 389-393

Gupta, S., Barrett, T., Whitmarsh, A.J., Cavanagh, J., Sluss, H.K., Derijard, B., and Davis, R. 1996. Selective interaction of JNK protein kinase isoforms with transcription factors. *EMBO J.* **15** : 2760-2770

Guyton, K.Z., Liu, Y.S., Gorospe, M., Xu, Q.B., and Holbrook, N.J. (1996). Activation of mitogen-activated protein kinase by H2O2-role in cell survival following oxidant injury. *J. Biol. Chem.* **271**: 4138-4142

Hagemann, C. and Rapp, U.R. 1999. Isotype-specific functions of Raf kinases. *Exp. Cell. Res.* **253**: 34-46

Ham, J., Babij, C., Whitfield, J., Pfarr, C.M., Lallemand, D., Yaniv, M. and Rubin, L.L. 1995. A c-Jun dominant negative mutant protects sympathetic neurons against programmed cell death. *Neuron* **14**: 927-939

Hardwick, M.J. 1997. Virus-induced apoptosis. *Adv. Pharmacol.* **41**: 295-336

Hardwick, J.M., Ketner, G. and Clem, R.J. 1998. Viral genes that modulate apoptosis, in: *Apoptosis genes*, J.W. Wilson, C. Booth, C.S. Potten, eds., Kluwer Acad. Publishers, pp. 243-279

Hardwicke, M.A. and Sandri-Goldin, R.M. 1994. The herpes simplex virus regulatory protein ICP27 contributes to the decrease in cellular mRNA levels during infection. *J. Virol.* **68**: 4797-4810

Hardwicke, M.A. and Schaffer, P.A. 1997. Differential effects of nerve growth factor and dexamethasone on herpes simplex virus type 1 oriL- and ori-S-dependent DNA replication in PC12 cells. *J. Virol.* **71**: 3580-3587

Hata, S., Koyama, A.H., Shiota, H., Adachi, A., Goshima, F., and Nishiyama, Y. 1999. Anti-apoptotic activity of herpes simplex virus type 2: the role of US3 protein kinase gene. *Microbes. Inf.* **1**: 601-607

Hayashi, Y., Iwasaka, T., Smith, C.C., Aurelian, L., Lewis, G.K., Ts'o, P.O.P.1985. Multistep transformation by defined fragments of herpes simplex virus type 2 DNA : oncogenic region and its gene product. *Proc. Natl. Acad. Sci. USA* **82**: 8493-8497

Hayashi, T., Sakai, K.-i., Sasaki, C., Itoyama, Y., Abe, K. 2000. Loss of Bag-1 immunoreactivity inrat brain after transient middle cerebral artery occlusion. *Brain Res.* **852**: 496-500

Haydar, T.F., Nowakowsky, R.S., Yarowsky, P.J., and Krueger, B.K. 2000. Role of founder cell deficit and delayed neurogenesis in microencephaly of the trisomy 16 mouse. *J. Neurosci.* **20**: 4156-4164

He, B., Gross, M. and Roizman, B. 1997. The γ(1)34.5 protein of herpes simplex virus type 1 complexes with protein phosphatase 1α to dephosphorylate the α subunit of the eukaryotic translation initiation factor 2 and preclude the shutoff of protein synthesis by double-stranded RNAactivated protein kinase. *Proc. Natl. Acad. Sci. USA* **94**:843-848

Hefti, F., Hartikka, J. and Knusel, B. 1989. Function of neurotrophic factors in the adult aging brain and their possible use in the treatment of neurodegenerative disease. *Neurobiol. Aging* **10**: 515-533

Hefti, F. and Knusel, B. 1990. Neurotrophic factors and neurodegenerative diseases, in *Growth factors and Alzheimer's disease*, F. Hefti, P. Brachet, B. Will and Y. Christen, eds., Springer-Verlag, Berlin-Heidelberg-New York, pp.1-13

Hemmings, B.A. 1997. Akt signaling: linking membrane events to life and death decisions. *Science* **275**: 628-630

Henderson, S., Rowe, M., Gregory, C., Croom-Carter, D., Wang, F., Longnecker R, Kieff, E, Rickinson, A. 1991. Induction of Bcl-2 expression by Epstein-Barr virus latent membrane protein 1 protects infected B cells from programmed cell death. *Cell* **65**: 1107-1115

Hengartner, M.O. 2000. The biochemistry of apoptosis. *Nature* **407** : 770-776

Herdegen, T., Kovary, K., Buhl, A., Bavo, R., Zimmerman, M. and Gass, P. 1995. Basal expression of the inducible factors c-Jun, JunB, JunD, c-Fos, FosB, nd Krox-24 in the adult rat brain. *J. Comp. Neurol.* **35**: 39-56

Herold, B.C., WuDunn, D., Soltys, N., and Spear, P.G. 1991. Glycoprotein C of herpes simplex virus type 1 plays a principal role in the adsorption of virus to cells and in infectivity. *J. Virol.* **65**: 1090-1098

Herold, B.C., Visalli, R.J., Sumarski, N., Brandt, C., and Spear, P.G. 1994. Glycoprotein C-independent binding of herpes simplex virus to cells requires cell surface heparan sulfate and glycoprotein B. *J. Gen. Virol.* **75**: 1211-1222

Highlander, S.L., Sutherland, S.L., Gage, P.J., Johnson, D.C., Levine, M., and Glorioso, J.C. 1987. Neutralizing monoclonal antibodies specific for herpes simplex virus glycoprotein D inhibit virus penetration. *J. Virol.* **61**: 3356-3364

Hilmas, C., Pereira, E.F.R., Alkondon, M., Rassoulpour, A., Schwarcz, R. and Albuquerque, E.X. 2001. The brain metabolite kynurenic acid inhibits α7 nicotinic receptor activity and increases non-α7 nicotinic receptor expression: physiopathological implications. *J. Neurosci. 21*: 7463-7473

Hirt, N. 1967. Selective extraction of polyoma DNA from infected mouse cell culture. *J.Mol.Biol*. **26**: 365-369

Ho, D.Y., Fink, S.L., Lawrence, M.S., Meier, T.J., Saydam, T.C., Dash, R., Sapolsky, R.M. 1995. Herpes simplex virus vector system: analysis of its in vivo and in vitro cytopathic effects. *J. Neurosci. Meth*. **57**: 205-215

Honess, R.W. and Roizman, R. 1974. Regulation of Herpesvirus macromolecular synthesis. I. Cascade regulation of the synthesis of three groups of viral proteins. *J.Virol*. **14**: 8-19

Honig, L.S., Rosenberg, R.N. 2000. Apoptosis and neurological disease. *Am. J. Med* **108**: 317-330

Hu, Y., Benedict, M.A., Wu, D., Inohara, N., Nunez, G. 1998. Bcl-xL interacts with Apaf-1 and inhibits Apaf-1-dependent caspase-9 activation. *Proc.Natl.Acad.Sci. USA* **95**: 4386-4391

Hu, B.R., Liu, C.L., Ouyang, Y., Blomgren, K., Siesjo, B.K.. 2000. Involvement of caspase-3 in cell death after hypoxia-ischemia declines during brain maturation. *J. Cereb. Blood Flow Metab*. **9**: 1294-1230

Huber, M., Watson, K.A., Selinka, H.C., Carthy, C.M., Klingel, K., McManus, B.M., and Kandolf, R. 1999 Cleavage of RasGAP and phosphorylation of mitogen-activated protein kinase in the course of coxsackievirus B3 replication. *J.Virol*. **73**: 3587-3594

Hudson, S.J., Dix, R.D. and Streilein, J.W. 1991. Induction of encephalitis in SJL mice by intranasal infection with herpes simplex virus type 1: a possible model of herpes simplex encephalitis in humans. *J. Infect. Dis*. **163**: 720-727

Hunter, J.C.R., Smith, C.C., Aurelian, L. 1995. The HSV-2 LA-1 oncoprotein is a member of a novel family of serine/threonine receptor kinases. *Int.J.Oncol.* **7**: 515-522

Hunter, J.C.R., Smith, C.C., Bose, D., Kulka, M., Broderick, R. and Aurelian, L. 1995. Intracellular internalization and signaling pathways triggered by the large subunit of HSV-2 ribonucleotide reductase (ICP10). *Virology* **210**: 345-360

Husi, H., Ward, M.A., Choudhary, J.S., Blackstock, W.P., Grant, S.G. 2000. Proteomic analysis of NMDA receptor-adhesion protein signaling complexes. *Nat. Neurosci.* **3**: 661-669

Hyman, B.T., Van Horsen, G.W., Damasio, A.R., Barnes, C.L. 1984. Alzheimer's disease: cell-specific pathology includes the hippocampal formation. *Science* **225**: 1168-1170

Iida, N., Namikawa, K., Kiyama, H., Ueno, H., Nakamura, S. and Hattoru, S. 2001. Requirement of Ras for the activation of mitogen-activated protein kinase by calcium influx, cAMP, and neurotrophin in hippocampal neurons. *J. Neurosci* .**21**: 6459-6466

Ikonomidou, C., Bosch, F., Miksa, M., Bittigau, P., Vockler, J., Dikranian, K., Tenkova, T.I., Stefovska, V., Turski, L. and Olney, J.W. 1999. Blockade of NMDA receptors and apoptotic neurodegeneration in the developing brain. *Science* **283**: 70-74

Improta-Brears, T., Whorton, A.R., Codazzi, F., York, J.D., Meyer, T., and McDonnel, D.P. 1999. Estrogen-induced activation of mitogen-activated protein kinase requires mobilization of intracellular calcium. *Proc. Natl. Acad. Sci. USA* **96** : 4686-4691

Ingemarson, R. and Lankinen, H. 1987. The herpes simplex virus type 1 ribonucleotide reductase is a tight complex of the type α2β2 composed of 40K and 140 K proteins, of which the latter shows multiple forms due to proteolysis. *Virology* **156**: 417-422

Irie, K., Gotoh, Y., Yashar, B.M., Errede, B., Nishida, E. and Matsumoto, K. 1994. Stimulatory effects of yeast and mammalian 14-3-3 proteins on the Raf protein kinase. *Science* **265**: 1716-1719

Ishitani, R., Sunaga, K., Tanaka, M., Aishita, H., Chuang, D.-M. 1997. Overexpression of glyceraldehyde-3-phosphate dehydrogenase is involved in low K^+-induced apoptosis but not necrosis of cultured cerebellar granule cells. *Mol. Pharmacol.* **51**: 542-550

Israeli, D., Tessler, E., Haupt, Y., Elkeles, A., Wilder, S., Amson, R., Telerman, A., Oren, M. 1997. A novel p53-inducible gene, PAG608, encodes a nuclear zinc finger protein whose overexpression promotes apoptosis. *EMBO J.* **16**: 4384-4392

Itoh, K., Brackenbury, R., Akeson, R. 1995. Induction of L1mRNA in PC12 cells by NGF is modulated by cell-cell contact and does not require the high affinity NGF receptor. *J. Neurosci.* **15**: 2504-2512

Ivins, K.J., Ivins, J.K., Sharp, J.P., and Cotman, C.W. 1999. Multiple pathways of apoptosis in PC12 cells. *J.Biol.Chem.* **274**: 2107-2112

Jaattela, M., Wissing, D., Kokholm, K., Kallunki, T., and Egeblad, M. 1998. Hsp70 exerts its anti-apoptotic function downstream of caspase-3-like proteases. *EMBO J.* **17**: 6124-6134

Jaattela, M. 1999. Escaping cell death: survival proteins in cancer. *Exp. Cell Res.* **248**: 30-43

Jacobson, J.G., Leib, D.A., Goldstein, D.J., Bogard, C.L., Schaffer, P.A., Weller, S.K., and Coens, D.M. 1989. A herpes simplex virus ribonucleotide reductase deletion mutant is defective for productive acute and reactivatable latent infections of mice and for replication in mouse cells. *Virology* **173**: 276-283

Jacobson, M.D., Burne, J.F., and Raff, M.C. 1994. Programmed cell death and Bcl-2 protection in the absence of a nucleus. *EMBO J.* **13** : 1899-1910

Jacobson, M.D., Weil, M., and Raff, M.C. 1996. Role of Ced-3-ICE-family proteases in staurosporine-induced programmed cell death. *J.Cell.Biol.* **133**: 1041-1051

Jacobson, J.G., Chen, S.H., Cook, W.J., Kramer, M.F., and Coen, D.M. 1998. Importance of the herpes simplex virus UL24 gene for productive ganglionic infection in mice. *Virology* **242**: 161-169

Jacque, J.M., Mann, A., Enslen, H., Sharuva, N., Brichacek, B., Davis, R.J., and Stevenson, M. 1998 Modulation of HIV-1 infectivity by MAPK, a virion associated kinase. *EMBO J.* **17**: 2607-2618

Janicke, R.U., Ng, P., Sprengart, M.L., Porter, A.G. 1998. Caspase-3 is required for alpha-fodrin cleavage but dispensable for cleavage of other death substrates in apoptosis. *J.Biol.Chem.* **273**: 15540-15545

Jariwalla, R.J., Aurelian, L., Ts'o, P.O.P. 1980. Tumorigenic transformation induced by a specific fragment of herpes simplex virus type 2 DNA. *Proc. Natl. Acad. Sci. USA* **77**: 2279-2283

Jerome, K.R., Fox, R., Chen, Z., Sears, A.E., Lee, H.-Y. and Corey, L .1999. Herpes simplex virus inhibits apoptosis through the action of two genes, US5 and US3. *J. Virol.* **73**: 8950-8957

Jerusalinsky, D., Kornisiuk, E., Izquierdo, I. 1997. Cholinergic neurotransmission and synaptic plasticity concerning memory processing. *Neurochem. Res.* **22**: 507-515

Jessel, T.M. and Goodman, C.S. 1996. Development: neural development: are there any surprises left? *Curr. Opin. Neurobiol.* **6**: 1-2

Johnson, D.C., Burke, R.L. and Gregory, T. 1990. Soluble forms of herpes simplex virus glycoprotein D bind to a limited number of cell surface receptors and inhibit virus entry into cells. *J. Virol.* **64**: 2564-2576

Johnson, P.A., Yoshida, K., Gage, F.H., and Friedman, T. 1992. Effects of gene transfer into cultured CNS neurons with a replication-defective herpes simplex virus type 1 vector. *Mol. Brain Res.* **12**: 95-102

Johnson. E.M. Jr, Cornbrooks, E.B., Deckwerth, T.L., Estus, S., Franklin, J.L., Freeman, R.S., Horigome, K., and Lampe, P.A. 1993. Molecular mechanism of programmed cell death in the developing nervous system, in *Neuronal Cell Death and Repair*, A.C. Cuello, ed., Elsevier Science Publishers, pp 23-31

Johnson Webb, S., Harrison, D.J., and Wyllie, A.H. 1997. Apoptosis: an overview of the process and its relevance in disease. *Adv. Pharmacol.* **41**: 1-31

Jones, C.A., Taylor, T. and Knipe, D.M. 2000. Biological properties of Herpes Simplex Virus Type 2 replication-defective mutant strains in a murine nasal infection model. *Virology* **278**: 137-150

Jordan, J., Galindo, M.F., Prehn, J.H.M., Weichselbaum, R.R., Beckett, M., Ghadge, G.D., Roos, R.P., Leiden, J.M., and Miller, R.J. 1997. p53 expression induces apoptosis in hippocampal pyramidal neuron cultures. *J. Neurosci.* **17**: 1397-1405

Jung, J.U. and Desrosiers, R.C.. 1995 Association of viral oncoprotein STP-C488 with cellular ras. *Mol. Cell. Biol.* **15**: 6506-6512

Junying, Y. and Yankner, B.A. 2000. Apoptosis in the nervous system. *Nature* **407**: 802-809

Kane, L. P., Shapiro, P., Stokoe, V.S., and Weiss, D. 1999. Induction of NF-[kappa]B by the Akt/PKB kinase. *Curr. Biol.* **9**: 601-604

Kaplan, D.R. and Miller, F.D. 2000. Neurotrophin signal transduction in the nervous system. *Curr.Opinion Neurobiol.* **10**: 381-391

Kastan, M.B., Zhan, Q., el-Deiry, W.S., Carrier, F., Jacks, T., Walsh, W.V., Plunkett, B.S., Vogelstein, B., Fornace, A.J. Jr. 1992. A mammalian cell cycle checkpoint pathway utilizing p53 and GADD45 is defective in ataxia telangiectasia. *Cell* **71**: 587-597

Kawanishi, M. 1997. Expression of Epstein Barr virus latent membrane protein 1 protects Jurkat cells from apoptosis induced by serum deprivation. *Virology* **228**: 244-250

Kelekar, A., Chang, B.S., Harlan, J.E., Fesik, S.W., Thomson, C.B. 1997. Bad is a BH3 domain-containing protein that forms an inactivating dimer with Bcl-xL. *Mol. Cell. Biol.* **17**: 7040-7046

Kibler, K.V., Shors, T., Perkins, K.B., Zeman, C.C., Banaszak, M.P., Biesterfeldt,J., Langland J.O., Jacobs, B.L. 1997. Double-stranded RNA is a trigger for apoptosis in vaccinia virus-infected cells. *J. Virol.* **71**: 1992-2003

Kiedrowski, L. 1998. The fifference between mechanisms of kainate and glutamate excitotoxicity *in vitro*: osmotic lesions versus mitochondrial depolarization. *Restor. Neurol. Neurosci.* **12**: 71-79

Kihara, T., Shimohama, S., Sawada, H., Honda, K., Nakamizo, T., Shibasaki, H., Kume, T., and Akaike, A. 2001. α7 nicotinic receptor transduces signals to phosphatidylinsitl 3-kinase to block a β-amyloid-induced neurotoxicity. *J. Biol. Chem.* **276**: 13541-13546

King, C.S., Cooper, J.A., Moss, B. and Twardzik, D.R. 1986. Vaccinia virus growth factor stimulates tyrosine protein kinase activity of A431 cell epidermal growth factor receptors. *Mol. Cell. Biol.* **6**: 332-336

Kitamura, Y., Shimohama, S., Kamoshima, W., Ota, T., Matsuoka, Y., Nomura, Y., Smith, M.A,. Perry, G., Whitehouse, P.J., Taniguchi, T. 1998. Alterations of proteins regulating apoptosis, Bcl-2, Bcl-X, Bax, Bak, Bad, ICH-1 and CPP32, in Alzheimer's disease. *Brain Res*. **780**: 260-269

Klein, N.P. and Schneider, R.J. 1997.Activation of Src family kinases by hepatitis B virus Hbx protein and coupled signaling to Ras. *Mol. Cell. Biol*. **17**: 6427-6436

Klein, N.P., Bouchard, M.J., Wang, L-H., Kobarg, C., and Schneider, R.J. 1999. Src kinases involved in hepatitis B replication. *EMBO J.* **18**: 5019-5027

Kokuba, H., Aurelian, L., and Burnett, J.W.1999. Herpes Simplex Virus associated erythema multiforme (HAEM) is mechanistically distinct from drug-induced erythema multiforme: interferon-gamma is expressed in HAEM lesions and tumor necrosis-alpha in drug-induced erythema multiforme lesions. *J. Invest. Dermatol*. **113**: 808-815

Koo, P.H., Liebl, D.J. 1992. Inhibition of nerve growth factor-stimulated neurite outgrowth by methylamine-modified alpha 2-macroglobulin. *J. Neurosci. Res*. **31**: 678-692

Korskinen, P., Sistonen, L., Evan, G., Morimoto, R. and Alitalo, K. 1991. Nuclear colocalization of cellular and viral myc proteins with HSP70 in myc-overexpressing cells. *J. Virol*. **65**: 842-851

Korsmeyer, S.J. 1992. Bcl-2: a repressor of lymphocyte death. *Immunol. Today* **13**: 285-288

Koyama, A.H., Akari, H., Adachi, A., Goshima, F. and Nishiyama, Y. 1998. Induction of apoptsis in Hep-2 cells by infection with herpes simplex virus type 2. *Arch. Virol.* **143**: 2435-2441

Koyama, A.H., Fukomori, T., Fujita, M., Irie, H., Adachi, A. 2000. Physiological significance of apoptosis in animal virus infection. *Microbes Inf.* **2**: 1111-1117

Krakauer, D.C. and Payne, R.J.H. 1997. The evolution of virus-induced apoptosis. *Proc. R. Soc. Lond.* **264**: 1757-1762

Krammer, P.H. 2000. CD95's deadly mission in the immune system. *Nature* **407**: 789-795

Kristie, T.M., Vogel, J.L. and Sears, A.E. 1999. Nuclear localization of the C1 factor (host cell factor) in sensory neurons correlates with reactivation of herpes simplex virus from latency. *Proc. Natl. Acad. Sci. USA* **96**: 1229-1233

Kristensson, K., Nennesmo, I., Persson, L., and Lycke, E. 1982. Neuron to neuron transmission of Herpes Simplex Virus. *J. Neurol. Sci.* **54**: 149-156

Krzanowski, J.J. 1995. AP-1 transcription factor complexes in CNS disorders and development. *J. Fla. Med. Ass.* **82**: 551-554

Kuan, C.Y., Yang, D.D., Samanta Roy, D.R., Davis, R.J., Rakic, P., Flavell, R.A. 1999. The Jnk1 and Jnk2 protein kinases are required for regional specific apoptosis during early brain development. *Neuron* **22**: 667-676

Kurino, M., Fukunaga, G., Ushio, Y. and Miyamoto, E. 1995. Activation of mitogen activated protein kinase in cultured rat hippocampal neurons by stimulation of glutamate receptors. *J. Neurochem.* **65**: 1282-1289

Kyriakis, J.M. and Avruch, J. 1996. Sounding the alarm: protein kinase cascades activated by stress and inflammation. *J. Biol. Chem.* **271**: 24313-24316

Kyriakis, J.M., Woodgett, J.R., and Avruch, J. 1995. The stress-activated protein kinases. A novel ERK subfamily responsive to cellular stress and inflammatory cytokines. *Ann. NY Acad. Sci.* **766**: 303-319

Lackey, K., Cory, M., Davis, R., Frye, S.V., Harris, P.A., Hunter, R.N., Jung, D.K., McDonald, O.B., McNutt, R.W., Peel, M.R., Rutkowske, R.D., Veal, J.M., and Wood, E.R. 2000. The discovery of potent cRaf1 kinase inhibitors. *Bioorg. Med. Chem. Lett.* **10**: 223-226

LaFerla, F.M. and Gilbert, J. 1997. β-amyloid induced neuronal cell death in transgenic mice and Alzheimer's disease, in *Neuromethods, vol. 29: Apoptosis Techniques and Protocols*, J. Pirier, ed., Humana Press, pp. 183-195

Laherty, C.D., Hu, H.M., Opipari, A.W., Wang, F., Dixit, V.M. 1992. The Epstein Barr virus LMP1 gene product induces A20 zinc finger protein expression by activating nuclear factor kappa B. *J. Biol. Chem.* **267**: 24157-24160

Lali, F.V., Hunt, A.E., Turner, S.J., and Foxwell, B.M.J. 2000. The pyridinyl imidazole inhibitor SB203580 blocks phosphoinositide-dependent protein kinase activity, protein kinase B phosphorylation, and retinoblastoma hyperphosphorylation in interleukin-2-stimulated T cells independently of p38 mitogen-activated protein kinase. *J. Biol. Chem.* **275**: 7395-7402

Lam, Q., Smibert, C.A., Koop, K.E., Lavery, C., Capone, J.P., Weinheimer, S.P., Smiley, J.R. 1996. Herpes simplex virus VP16 rescues viral mRNA from destruction by the virion host shutoff function. *EMBO J.* **15**: 2575-2581

Lapchak, P.A., Jiao, S.S., Miller, P.J., Williams, L.R., Cummins, V., Inouye, G., Matheson, C.R. and Yan, Q. 1996. Pharmacological characterization of glial cell line-derived neurotrophic factor: implications for GDNF as a therapeutic molecule to treat neurodegenerative diseases. *Cell Tissue Res.* **286**: 179-189

Lazebnik, Y.A., Kaufmann, S.H., Desnoyers, S., Poirier, G.G., Earnshaw, W,C. 1994. Cleavage of poly (ADP-ribose) polymerase by a proteinase with properties like ICE. *Nature* **371** : 346-347

Lee, Y.I., Kang-Park, S., Do, S.-I., and Lee, Y.i. 2001. The hepatitis B virus-X protein activates a phosphatidylinositol 3-kinase-dependent survival signaling cascade. *J. Biol. Chem.* **276**: 16969-16977

Leist, M., Gantner, F., Bohlinger, I., Tiegs, G., Germann, P.G., Wendel, A. 1995. Tumor necrosis factor-induced hepatocyte apoptosis precedes liver failure in experimental murine shock models. *Am. J. Pathol.* **146**: 1220-1234

Leist, M. and Nicotera, P. 1998. Apoptosis versus necrosis: the shape of neuronal cell death, in *Apoptosis: mechanisms and role in disease*, S. Kumar, ed., Berlin-New York; Springer, pp. 105-135

Leopardi, R., and Roizman B. 1996. The herpes simplex virus major regulatory protein ICP4 blocks apoptosis induced by the virus or by hyperthermia. *Proc. Natl. Acad. Sci. USA* **93**: 9583 - 9587

Leopardi, R., Van Sant, C. and Roizman, B. 1997. The herpes simplex virus 1 protein kinase US3 is required for protection from apoptosis induced by the virus. *Proc. Natl. Acad. Sci. USA* **94**: 7891 - 7896

Lens, D., Dyer, M.J.S., Garcia Marco, J.M., De Schouwer, P.J., Hamoudi, R.A., Jones, D., Farahat, N., Matutes, E., Catowsky, D. 1997. p53 abnormalities in CLL are associated with excess of prolymphocytes and poor prognosis. *Br. J. Haematol*. **99**: 848-857

LeRoith, D., Roberts, C., Werner, H., Bondy, C., Raizada, M., Adamo, M. 1993. Insulin-like growth factors in the brain, in *Neurotrophic factors*, Loughlin S, Fallon J, eds., Boston, Acad. Press, pp. 391-442

Levi-Montalcini, R., and Booker, B. 1960. Destruction of sympathetic ganglia in mammals by an antiserum to the nerve-growth promoting factor. *Proc. Natl. Acad. Sci. USA* **46**: 384-391

Levine, B., Huang, Q., Isaacs, J.T., Reed, J.C., Griffin, D.E. and Hardwick, J.M. 1993. *Nature* (London) **361**: 739-742

Levine, E., Dreyfus, C., Black, I., Plummer, M. 1995. Differential effects of NGF and BDNF on voltage-gated calcium currents in embryonic basal forebrain neurons. *J. Neurosci.* **15**: 3084-3091

Li, H., Zhu, H., Xu, C.-J., and Yuan, J. 1998. Cleavage of BID by caspase-8 mediates the mitochondrial damage in the Fas pathway of apoptosis. *Cell* **94**: 491-501

Li, M., Ona, V.O., Guegan, C. and Chen, M. 2000. Functional role of caspase-1 and caspase-3 in an ALS transgenic mouse model. *Science* **288**: 335-339

Lipton, P. 1999. Ischemic cell death in brain neurons. *Physiol. Reviews* **79**: 1431-1568

Lu, B., Figurov, B. 1997. Role of neurotrophins in synapse development and plasticity. *Rev. Neurosci.* **8**: 1-12

Luo, J. and Aurelian, L. 1992. The transmembrane helical segment but not the invariant lysine is required for the kinase activity of the large subunit of herpes simplex virus type 2 ribonucleotide reductase (ICP10). *J. Biol. Chem.* **267**: 9645-9653

Lynas, C., Hill, T.J., Maitland, N.J., and Love, S. 1993. Latent infection with the MS strain of herpes simplex virus type 2 in the mouse following intracerebral inoculation. *J. Neurol. Sci.* **120**: 107-114

Macen, J.L., Upton, C., Nation, N., McFadden, G. 1993. SERP1, a serine proteinase inhibitor encoded by myxoma virus, is a secreted glycoprotein that interferes with inflammation. *Virology* **195**: 348-363

Malek, A.M., Goss, G.G., Jiang, L., Izumo, S., Alper, S.L. 1998. Mannitol at clinical concentrations activates multiple signaling pathways and induces apoptosis in endothelial cells. *Stroke* **29**: 2631-2640

Manji, G.A., Hozak, R.R., LaCount, D.J., Friesen, P.D. 1997. Baculovirus inhibitor of apoptosis functions at or upstream of the apoptotic suppressor P35 to prevent programmed cell death. *J. Virol.* **71**: 4509-4516

Marais, R., Wynne, J., Treisman, R. 1993. The SRF accessory protein Elk-1 contains a growth factor-regulated transcriptional activation domain. *Cell* **73**: 381-393

Marcellus, R.C., Teodoro, J.G., Wu, T., Brough, D.E., Ketner, G., Shore, G.C., Branton, P.E. 1996. Adenovirus type 5 early region 4 is responsible for E1A-induced p53-independent apoptosis. *J.Virol.* **70**: 6207-6215

Margolis, T.P., Sedarati, F., Dobson, A.T., Feldman, L.T., and Stevens, J.G. 1992. Pathways of viral gene expression during acute neuronal infection with HSV-1. *Virology* **189**: 150-160

Marsters, S.A., Ayres, T.M., Skubatch, M., Gray, C.L., Rothe, M., and Ashkenazi, A. 1997. Herpesvirus entry mediator, a member of the tumor necrosis factor receptor (TNFR) family, interacts with members of the TNFR-associated factor family and activates the transcription factors NF-kB and AP-1. *J. Biol. Chem.* **272**: 14029-14032

Martin, G.M. 1996. Intrinsic biological aging as underlying pathogenetic mechanisms in dementias of the Alzheimer's type, in *Alzheimer's disease cause(s), diagnosis, treatment and care*, ed. by Z.S. Khachaturian and T.S. Radebaugh, CRC Press

Masquilier, D. and Sassone-Corsi, P. 1992. Transcriptional cross-talk: nuclear factors CREM and CREB bind to AP-1 sites and inhibit Activation by Jun. *J. Biol. Chem* **267**: 22460-22466

Matsuzawa, S.-i., Takayama, S., Froesch ,B.A., Zapata, J.M., and Reed, J.C. 1998. p53-inducible human homologue of Drosophila seven in absentia (Siah) inhibits cell growth: suppression by BAG-1. *EMBO J.* **17**: 2736-2747

Maundrell, K., Antonsson, B., Magnenat, E., Camps, M., Muda, M., Chabert, C., Gillieron, C., Boschert, U., Vial-Knecht, E., Martinou, J.C. and Arkinstall, S. 1997. Bcl-2 undergoes phosphorylation by c-jun N-terminal kinase/Stress-activated protein kinases in the presence of the constitutively active GTP-binding protein Rac1. *J. Biol. Chem.* **272**: 25238-25242

May, W.S., Tyler, P.G., Ito, T., Armstrong, D.K., Qatsha, K.A., Davidson, N.E. 1994. Interleukin-3 and bryostatin-1 mediate hyperphosphorylation of BCL2alpha in association with suppression of apoptosis. *J. Biol. Chem*. **269**: 26865-26870

Mazzoni, I.E., Said, F.A., Aloyz, R., Miller, F.D., Kaplan, D. 1999. Ras regulates sympathetic neuron survival by suppressing the p53-mediated cell death pathway. *J. Neurosci*. **19**: 9716-9727

McGeoch, D.J. 1990. Evolutionary relationships of virion glycoprotein genes in the S regions of alphavirus genomes. *J.Gen.Virol*. **71**: 2361-2367

McLean, T.I. and Bachenheimer, S.L.1999. Activation of c-Jun N-terminal kinase by herpes simplex virus type 1 enhances viral replication. *J. Virol*. **73**: 8415-8426

Meyding-Lamadé, U., Lamadé, W., Heβ, T., Gosztonyi, G., Kehm, R., Sartor, K., and Hacke, W. 1996. A mouse model of herpes simplex virus encephalitis: diagnostic brain imaging by magnetic resonance imaging. *J. Int. In Vivo Res*. **10**: 563-568

Meyding-Lamadé, U., Haas, J., Lamadé, W., Stingele, K., Kehm, R., Fath, A., Heinrich, K., Hagenlocher, B.S., Wildemann, B. 1998. Herpes Simplex Virus encephalitis: long-term comparative study of viral load and the expression of immunologic nitric oxide synthase in mouse brain tissue. *Neurosci. Lett*. **244**: 9-12

Miyashita, T. and Reed, J.C. 1995. Tumor suppressor p53 is a direct transcriptional activator of the human bax gene. *Cell* **80**: 293-299

Mignotte, B. and Vayssiere, J.-L. 1998. Mitochondria and apoptosis. *Eur. J. Biochem.* **252**: 1-15

Minden, A., Karin, M. 1997. Regulation and function of the JNK subgroup of MAP kinases. *Bioch. Bioph. Acta* **1333** : F85-F104

Mobley, W.C., Neve, R.I., Prusiner, S.B., McKinley, M.P. 1988. Nerve growth factor induces gene expression for prion- and Alzheimer's beta-amyloid proteins. *Proc. Natl. Acad. Sci. USA* **85**: 9811-9815

Morishima, Y., Gotoh, Y., Zieg, J., Barrett, T., Takano, H., Flavell, R., Davis, R.J., Shirasaki, Y., and Greenberg, M.E. 2001. Beta-amyloid induces neuronal apoptosis via a mechanism that involves the c-jun N-terminal kinase pathway and the induction of Fas ligand. *J. Neurosci.* **21**: 7551-7560

Morrison, R. 1990. The Raf-1 kinase as a transducer of mitogenic signals. *Cancer Cells* **2**: 377-382

Morrison, R. 1993. Epidermal growth factor: structure, expression and functions in the central nervous system, in *Neurotrophic Factors*, Loughlin S., Fallon J. eds, Boston, Academic Press, pp. 339-358

Morrison, D.K., G. Heidecker, U.R. Rapp, and T.D. Copeland. 1993. Identification of the major phosphorylation sites of the Raf-1 kinase. *J. Biol. Chem.* **268**: 17309-17316

Mosser, D.D., Caron, A.W., Bourget, L., Denise-Larose, C., and Massie, B. 1997. Role of human heat shock protein hsp70 in protection against stress-induced apoptosis. *Mol. Cell. Biol.* **1**7: 5317-5327

Mufson, E.J., Bothwell, M., Hersh, L.B. and Kordower, J.H. 1989. Nerve growth factor receptor immunoreactive profiles in the normal aged human basal forebrain: colocalization with cholinergic neurons. *J. Comp. Neurol.* **285**: 196-217

Munger, J. and Roizman, B. 2001. The US3 protein kinase of herpes simplex virus type 1 mediates the posttranslational modification of BAD and prevents BAD-induced programmed cell death in the absence of other viral proteins. *Proc. Natl. Acad. Sci. USA* **98**: 10410-10415

Munger, J., Chee, A.V., and Roizman, B. 2001. The US3 protein kinase blocks apoptosis induced by the d120 mutant of herpes simplex virus type 1 at a premitochondrial stage. *J. Virol.* **75**: 5491-5497

Murphy, M. and Levine, A.J. 1998. The role of p53 in apoptosis, in *Apoptosis genes*, Wilson J.W., Booth C. and Potten C.S., eds., pp 5-35

Nakai, M., Qin, Z.-H., Chen, J.-F., Wang, Y., and Chase, T. 2000. Kainic acid-induced apoptosis in rat striatum is associated with nuclear factor-κB activation. *J. Neurochem.* **74**: 647-658

Nakayama, H., Numakawa, T., Ikeuchi, T., Hatanaka, H. 2001. Nicotine-induced phosphorylation of extracellelar signal-regulated protein kinase and CREB in PC12 cells. *J. Neurochem.* **79**: 489-498

Natsume, A., Mata, M., Goss, J., Huang, S., Wolfe, D., Oligino, T., Glorioso, J., and Fink, D.J. 2001. Bcl-2 and GDNF delivered by HSV-mediated gene transfer act additively to protect dopaminergic neurons from 6-OHDA-induced degeneration. *Exper. Neurol.* **169**: 231-238

Newcomb, R., Sun, X., Taylor, I., Curthoys, N., Giffard, R.G. 1997. Increased production of extracellular glutamate by the mitochondrial glutaminase following neuronal death. *J. Biol. Chem.* **272**: 11276-11282

Nicola, A.V., de Leon, M.P., Xu, R., Hou, W., Whitbeck, J.C., Krummenacher, C., Montgomery, R.I., Spear, P.G., Eisenberg, R.J., and Cohen, G.H. 1998. Monoclonal antibodies to distinct sites on herpes simplex virus (HSV) glycoprotein D block HSV binding to HVEM. *J. Virol.* **72**: 3595-3601

Nicholson, D.W., Ali, A., Thornberry, N.A., Vaillancourt, J.P., Ding, C.K., Gallant, M., Gareau, Y., Griffen, P.R., Labelle, M., Lazebnik , Y.A ., Munday, N.A., Raju, S.M., Smulson, M.E., Yamin, T.-T., Yu, V.L. and Miller DK. 1995. Identification and inhibition of the ICE/CED-3 protease necessary for mammalian apoptosis. *Nature* **376**: 37-43

Nicotera, P. and Leist, M. 1997. Energy supply and the shape of death in neurons and lymphoid cells. *Cell Death Differ.* **4**: 435-442

Nikas, I., McLauchlan, J., Davison, A.J., Taylor, W.R. and Clements, J.B. 1986. Structural features of ribonucleotide reductase. *Proteins: structure, function and genetics* **1**: 376-384

Olivier, N., Ali, C., Docagne, F., Plawinski, L., MacKenzie, E.T., Viven, D., and Buisson, A. 2001. Neuroprotection mediated by glial cell line-derived neurotrophic factor: invlvement of a reduction of NMDA-induced calcium influx by the mitogen-activated protein kinase pathway. *J. Neurosci.* **21**: 3024-3033

Oppenheim, R.W. 1991. Cell death during development of the nervous system. *Annu. Rev. Neurosci.* **14**: 453-501

Oshima ,M., Sithanandam, G., Rapp, U., Guroff, G. 1993. The phosphorylation and activation of B-Raf in PC12 cells stimulated by nerve growth factor. *J. Biol. Chem.* **266**: 23753-23760

Overton, H., McMillan, D., Hope, L., and Wong-Kai-In, P. 1994. Production of host shutoff-defective mutants of herpes simplex virus type 1 by inactivation of the UL13 gene. *Virology* **20**2: 97-106

Packham, G., Brimmell, M., and Cleveland, J.L. 1997. Mammalian cells express two differently localized Bag-1 isoforms generated by alternative translation initiation. *Biochem. J.* **328**: 807-813

Pasinelli, P., Borchelt, D.R., Houseweart, M.K., Cleveland, D.W., and Brown, R.H. Jr. 1998. Caspase-1 is activated in neural cells and tissue with amyotrophic lateral sclerosis-associated mutations in copper-zinc superoxide dismutase. *Proc. Natl. Acad. Sci. USA* **9**5: 15763-15768

Pasinelli, P., Hoseweart, M.K., Brown, R.H. Jr. and Cleveland, D.W. 2000. Caspase-1 and -3 are sequentially activated in motor neuron death in Cu, Zn superoxide dismutase-mediated familial amyotrophic lateral sclerosis. *Proc. Natl. Acad. Sci. USA* **97**: 13901-13906

Peng, T., Hunter, J.R.C. and Nelson, J.W.1996. The novel protein kinase of the R1 subunit of herpes simplex virus has autophosphorylation and transphosphorylation activity that differs in its ATP requirements for HSV-1 and HSV-2. *Virology* **216**: 184-196

Peng, T., Sadusky, T., Li, Y., Coulton, G.R., Zhang, H., Archard, L.C. 2001. Altered expression of Bag-1 in Coxsackievirus B3 infected mouse heart. *Cardiovascular Res* **50**: 46-55.

Pennypacker, K.R., Thai, L., Hong, J.-S., and McMillian, M.K. 1994. Prolonged expression of AP-1 transcription factors in the rat hippocampus after systemic kainate treatment. *J. Neurosci.* **14**: 3998-4006

Pennypacker, K.R., Hudson, P.M., Hong, J. and McMillan, M.K. 1995. DNA binding activity of CREB transcription factor during ontogeny of central nervous system. *Dev. Brain Res.* **86**: 242-249

Perlman, D. and Halverson, H.O. 1983. A putative signal peptide recognition site and sequence in eukaryotic and prokaryotic signal peptides. *J. Mol. Biol.* **167**: 391-409

Perng, G.-C., Jones, C., Ciacci-Zanella, J., Stone, M., Henderson, G., Yukht, A., Slanina, S.M., Hofman, F.M., Ghiasi, H., Nesburn, A.B, and Wechsler, S.L. 2000. Virus-induced neuronal apoptosis blocked by the herpes simplex virus latency-associated transcript. *Science* **287**:1500-1503

Perry, E.K., Morris, C.M., Court, J.A., Cheng, A., Fairbairn, A.F., McKeith, I.G., Irving, D., Brown, A., Perry, R.H. 1995. Alteration in nicotine binding sites in Parkinson's disease, Lewy body dementia and Alzheimer's disease: possible index of early neuropathology. *Neuroscience* **64**: 385-395

Philpott, K. L., McCarthy, M. J., Klippel, A. & Rubin, L. L. 1997. Activated phosphatidylinositol 3-kinase and Akt kinase promote survival of superior cervical neurons. *J. Cell. Biol.* **139**: 809-815

Pittman, R.N., Wang, S., DiBenedetto, A.J., and Mills, J.C.A. 1993. A system for characterizing cellular and molecular events in programmed cell death. *J.Neurosci.* **13**: 3669-3680

Plattner, R., Gupta, S., Khosravi-Far, S., Sato, K.Y., Perucho, M., Der, C.J., and Stanbridge, E.J. 1999. Differential contribution of the ERK and JNK mitogen-activated protein kinase cascades to Ras transformation of HT1080 fibrosarcoma and DLD-1 colon carcinoma cells. *Oncogene* **18**: 1807-1817

Plioplys, A.V. 1991. Trisomy 16 mouse model of Alzheimer's disease, in *Alzheimer's disease: basic mechanisms, diagnosis and therapeutic strategies*, Iqbal K., McLachlan D.R.C., Winblad R. and Wisniewski H.M. eds., John Wiley & Sons Ltd., pp 479-485

Portera-Cailliau, C., Hedreen, J.C., Price, D.L., Koliatsos, V.E. 1995. Evidence for apoptotic cell death in Huntington disease and excitotoxic animal models. *J. Neurosci.* **15**: 3775-3787

Portera-Cailliau, C., Price, D.L., and Martin, L.J. 1997. Excitotoxic neuronal death in the immature brain is an apoptosis-necrosis continuum. *J. Comp. Neurol.* **378**: 70-87

Portera-Cailliau, C., Price, D.L., and Martin, L.J. 1997. Non-NMDA and NMDA-receptor mediated excitotoxic neuronal cell deaths in adult brain are morphologically distinct:further evidence for an apoptosis-necrosis continuum. *J. Comp. Neurol.* **378**: 88-104

Preston, G.A., Lyon, T.T., Yin, Y., Lang, J.E., Solomon, G., Annab, L., Srinivasan, D.G., Alcorta, D.A. and Barrett, J.C. 1996. Induction of apoptosis by c-fos protein. *Mol. Cell. Biol.* **16**: 211-218

Prikhod'ko, E.A. and Miller, L.K. 1996. Induction of apoptosis by baculovirus transactivator IE-1. *J. Virol.* **70**: 7116-7124

Purifoy, D.J.M. and Powell, K.L. 1977. Interference between strains of type 1 and type2 herpes simplex virus. *Virology* **77**: 84-94

Raingeaud, J., Whitmarsh, A.J., Barrett, T., Derijard, B., Davis, R.J. 1996. MKK-3 and MKK6-regulated gene expression is mediated by the p38 mitogen-activated protein kinase signal transduction pathway. *Mol. Cell. Biol.* 16: 1247-1255

Reyes, G.R., LaFemina, R., Hayward, S.D., and Hayward, G.S.. 1979. Morphological transformation by DNA fragments of human herpesviruses : evidence for two distinct transforming regions in herpes simplex virus types 1 and 2 and lack of correlation with biochemical transfer of the thymidine kinase gene. *Cold Spring Harbor Symp. Quant. Biol.* **44**: 629-641

Riccio, A., Ahn, S., Davenport, C. M., Blendy, J. A. and Ginty, D. D. 1999. Mediation by a CREB family transcription factor of NGF-dependent survival of sympathetic neurons. *Science* **286**: 2358-2361

Richards, S.-J., Waters, J.J., Wischik, C., Sparkman, D.R., White, C.L. III, Beyreuther, K., Masters, C., Abraham, C.R. and Dunnett, S.B. 1991. A new model for studying the neuropathology of Alzheimer's disease derived from transplantation of trisomy 16 CNS tissues, in *Alzheimer's disease: basic mechanisms, diagnostic and therapeutic strategies*, Iqbal K, McLachlan R., Winblad R. and Wisniewski H.M. eds., John Wiley & Sons Ltd., pp. 478-497

Roberson, E.D., English, J.D., Adams, J.P., Selcher, J.C., Kondratick, C. and Sweatt, J.D. 1999. The mitogen-activated protein kinase cascade couples PKA and PKC to cAMP response element binding protein phosphorylation in area CA1 of hippocampus. *J. Neurosci.* **19**: 4337-4348

Robinson, M.J. and Cobb, M.H. 1997. Mitogen-activated protein kinase pathways. *Curr. Opin. Cell. Biol.* **9**: 180-186

Rodahl, E. and Haarr, L. 1997. Analysis of the 2-kilobase latency associated transcript expressed in PC12 cells productively infected with herpes simplex virus type 1: evidence for a stable, nonlinear structure. *J. Virol.* **71**: 1703-1707

Rodriguez-Tebar, A., Dechant, G., Gotz, R., Barde, Y.A. 1992. Binding of neurotrophin-3 to its neuronal receptors and interactions with nerve growth factor and brain-derived neurotrophic factor. *EMBO J.* **11**: 917-922

Roizman, B. and Sears, A.E. 1996. Herpes simplex viruses and their replication, in *Fields' Virology*, 3rd ed., ed. by B.N. Fields, D.M. Knipe, P.M. Howley et.al, Lippincott-Raven Publishers, PA , p. 2231-2295

Rosen, D.R., Siddique, T., Patterson, D., Figlewicz, D.A., Sapp, P., Hentati, A., Donaldson, D., Goto, J., O'Regan, J.P., Deng, H.X. et al. 1993. Mutations in Cu/Zn superoxide dismutase gene are associated with familial amyotrophic lateral sclerosis. *Nature* **362**: 59-62

Rosen, L.B., Ginty, DD., Weber, M.J., Greenberg, M.E. 1994. Membrane depolarization and calcium influx stimulate MEK and MAP kinase va activation of Ras. *Neuron* **12**: 1207-1221

Rosenblum, K., Futte,r M., Jones, M., Hulme, E.C. and Bliss, T.V.P. 2000. ERKI/II regulation by the muscarinic acetylcholine receptors in neurons. *J. Neurosci.* **2**0: 977-985

Roulston, A., Marcellus, R.C. and Branton, P.E. 1999. Viruses and apoptosis. *Annu. Rev. Microbiol.* **53**: 577-628

Rouquet, N., Allemand, I., Molina, T., Bennoun, M., Briand, P., Joulin, V. 1995. Fas-dependent apoptosis is impaired by SV40 T-antigen in transgenic liver. *Oncogene* **11**: 1061-1067

Roy, N., Deveraux, Q.L., Takahashi, R., Salvesen, G.S., Reed, J.C. 1997. The c-IAP-1 and c-IAP-2 proteins are direct inhibitors of specific caspases. *EMBO J.* **16**: 6914-6924

Ruther, U., Garber, C., Komitowski, D., Muller, R. and Wagner, E.F. 1987. Deregulated c-fos expression interferes with normal bone development in transgenic mice. *Nature* **325**: 412-416

Sakhi, S., Bruce, A., Sun, N., Tocco, G., Baudry, M., and Schreiber, S.S. 1994. P53 induction is associated with neuronal damage in the central nervous system. *Proc. Natl. Acad. Sci. USA* **91**: 7525-7529

Saldanha, J., Sutton, R.N.P., Gannicliffe, A., Faragher, B., Itzaki, R.F. 1986. Detection of HSV-1 DNA by in situ hybridisation in human brain after immunosuppression. *J. Neurol. Neurosurg. Psych* **49**: 613-619

Sasaki, M., Dawson, V.L., and Dawson, T.M. 2000. The NO signaling in the brain, in *Cerebral signal transduction*, Reith M.E.A., ed., Humana Press Inc., Totowa, NJ, pp. 151-173

Sasaoka, T., Draznin, B., Leitner, J.W., Langlois, W.J. and Olefsky, J.M. 1994a. *J. Biol. Chem.* **269**: 10734-10738

Sasaoka, T., Langloi,s W.J., Leitner, J.W., Draznin, B. and Olefsky, J.M. 1994b. *J. Biol. Chem.* **269**: 32621-32625

Savill, J. and Fadok, V. 2000. Corpse clearance defines the meaning of cell death. *Nature* **407**: 784-788

Sawa, A. 1999. Neuronal cell death in Down's syndrome. *J. Neural. Transm.* Suppl.**57**: 87-97

Scheffner, M., Werness, B.A., Huibregtse, J.M., Levine, A.J. and Howley, P.M. 1990. The E6 oncoprotein encoded by human papillomavirus types 16 and 18 promotes the degradation of p53. *Cell* **63**:1129-1136

Schlingenspien, K.-H., Wollnik, F., Kunst, M., Schlingenspien, R., Herdegen, T., and Brysch, W. 1994. The role of Jun transcription factor expression and phosphorylation in neuronal differentiation, neuronal cell death, and plastic adaptations in vivo. *Cell. Mol. Neurobiol.* **14**: 487-505

Schonthal, A., Buscher, M., Angel, P., Rahmsdorf, H.J., Ponta, H., Hattori, K., Chiu, R., Karin, M., Herrlich, P. 1989. The Fos and Jun/AP-1 proteins are involved in the downregulation of Fos transcription. *Oncogene* **4**: 629-636

Schotte, P., Declercq, W., Van Huffel, S., Vandenabeele, P., Beyaert, R. 1999. Non-specific effects of methyl ketone peptide inhibitors of caspases. *FEBS Lett.* **442**: 117-121

Schouten, G.J., Vertegaal, A.C.O., Whiteside, S.T., Israel, A., Toebes, M., Dorsman, J.C., van der Eb, A.J. Zantema, A. 1997. I kappa B alpha is a target for the mitogen-activated 90 kDa ribosomal S6 kinase. *EMBO J.* **16**: 3133-3144

Schreiber, M,. Sedger, L. and McFadden, G. 1997. Distinct domains of M-T2, the myxoma virus tumor necrosis factor (TNF) receptor homolog, mediate extracellular TNF binding and intracellular apoptosis inhibition. *J.Virol*. **71**: 2171-2181

Schubert, D. and Piasecki, D. 2001. Oxidative glutamate toxicity can be a component of the excitotoxicity cascade. *J. Neurochem*. **21**: 7456-7462

Schulz, J.B., Bremen, D., Reed, J.C., Lommatzsch, J., Takayama, S., Wullner, U., Loschmann, P.-A., Klockgether, T., and Weller, M. 1997. Cooperative interception of neuronal apoptosis by Bcl-2 and Bag-1 expression: prevention of caspase activation and reduced production of reactive oxygen species. *J. Neurochem*. **69**: 2075-2086

Schweighardt, B. and Atwood, W.J. 2001. Virus receptors in the human central nervous system. *J. Neurovirol*. **7**: 187-195

Segal, R.A. and Greenberg, M.E. 1996. Intracellular signaling pathways activated by neurotrophic factors. *Annu. Rev. Neurosci*. **19**: 463-489

Shepherd, G.M. 1994. Learning and memory, in *Neurobiology*, 3rd ed., Oxford University Press Inc., pp. 618-650

Shimamura, A., Ballif, B.A., Richards, S.A and Bleni,s J. 2000. Rsk 1 mediates a MEK-MAP kinase cell survival signal. *Curr. Biol*. **10**: 127-135

Shoji, M., Iwakami, N., Takeuchi, S., Waragai, M., Suzuki, M., Kanazawa, I., Lippa, C.F., Ono, S., Okazawa, H. 2000. JNK activation is associated with intracellular β-amyloid accumulation. *Mol. Brain Res*. **85**: 221-233

Sieber-Blum, M. 1991. Role of neurotrophic factor BDNF and NGF in the commitment of pluripotent neural crest cells. *Neuron* **6**: 949-955

Sieg, S., Yildirim, Z., Smith, D., Kayagaki, N., Yagita, H., Huang, Y., and Kaplan, D. 1996. Herpes simplex virus type 2 inhibition of Fas ligand expression. *J. Virol.* **70**: 8747-8751

Silva, A.J., Kogan, J.H., Frankland, P.W. and Kida, S. 1998. CREB and memory. *Annu. Rev. Neurosci.* **21**: 127-148

Simonian, N.A., Getz, R.L., Leveque, J.C., Konradi, C. and Coyle, J.T. 1996. Kainic acid induces apoptosis in neurons. *Neuroscience* **75**: 1047-1055

Smeal, T., Angel, P., Meek, J., Karin, M. 1989. Different requirements for formation of Jun: Jun and Jun: Fos complexes. *Genes Dev.* **3**: 2091-2100

Smeyne, R.J., Vendrell, M., Hayward, M., Baker, S.J., Miao, G.G, Schilling, K., Robertson, L.M., Curran, T., and Morgan, J.I. 1993. Continuous c-fos expression precedes programmed cell death in vivo. *Nature* **363**: 401-408

Smith, C.C., Wymer, J.P., Luo, J., Aurelian, L. 1991. Genomic sequences homologous to the protein kinase region of the bifunctional herpes simplex virus type 2 protein ICP10. *Virus Genes* **5**: 215-225

Smith, C.C., Kulka, M., Wymer, J.P., Chung, T.D., and Aurelian, L. 1992. Expression of the large subunit of herpes simplex virus type 2 ribonucleotide reductase (ICP10) is required for virus growth and neoplastic transformation. *J. Gen. Virol.* **73**: 1417-1428

Smith, C.C., Luo, J.H., Hunter, J.C.R., Ordonez, J.V., and Aurelian, L. 1994. The transmembrane domain of the large subunit of HSV-2 ribonucleotide reductase (ICP10) is required for protein kinase activity and transformation-related signaling pathways that result in ras activation. *Virology* **200**: 598-612

Smith, C.C. and Aurelian, L. 1997. The large subunit of herpes simplex virus type 2 ribonucleotide reductase (ICP10) is associated with the virion tegument and has PK activity. *Virology* **234**: 235-242

Smith, D.H., Chen, X.-H., Pierce, J.E.S., Wolf, J.A., Trojanowsky, J.Q., Graham, D.I., McIntosh, T.K. 1997. Progressive atrophy and neuron death for one year following brain trauma in the rat. *J. Neurotrauma* **14**: 715-727

Smith, C.C., Peng, T., Kulka, M. and Aurelian, L. 1998. The PK domain of the large subunit of herpes simplex virus type 2 ribonucleotide reductase (ICP10) is required for immediate-early gene expression and virus growth. *J.Virol.* **72**: 9131-9141

Smith, C.C., Nelson, J., Aurelian, L., Gober, M. and Goswami, B.B. 2000. Ras-GAP binding/ phosphorylation by HSV-2 RR1PK (ICP10) and activation of the Ras/MEK/MAPK mitogenic pathway are required for timely onset of virus growth. *J.Virol.* **74**: 10417-10429

Soh, J.-W., Lee, E.H., Prywes, R., and Weinstein, B. 1999. Novel roles of specific isoforms of protein kinase C in activation of the c-fos serum response element. *Mol. Cell. Biol.* **19**: 1313-1324

Sonnenberg, J.L., Mitchelmore, C., Macgregor-Leon, P.F., Hempstead, J., Morgan, J.I., Curran, T. 1989. Glutamate receptors agonists increase expression of Fos, fra, and AP-1 DNA binding activity in the mammalian brain. *J Neurosci. Res.* **24**:72-80

Sontag, E., Fedorov, S., Kamibayashi, C., Robbins, D., Cobb, M., and Mumby, M. 1993. The interaction of SV40 small tumor antigen with protein phosphatase 2A stimulates MAP kinase pathway and induces cell proliferation. *Cell* **75**: 887-897

Speck, P.G., and Simmons, A. 1991. Divergent molecular pathways of productive and latent infection with a virulent strain of Herpes simplex virus type 1. *J. Virol.* **65**: 4001-4005

Steiner, I., Kennedy, P.G . 1991.Herpes simplex virus latency in the nervous system: a new model. *Neuropathol. Appl .Neurobiol.* **17**: 433-440

Stephens, L., Anderson, K., Stokoe, D., Erdjument-Bromage, H., Painter, G.F., Holmes, A.B., Gaffney, P.R., Reese, C.B., McCormick, F., Tempst, P., Coadwell, J. and Hawkins, P.T. 1998. Protein kinase B kinases that mediate phosphatidylinositol 3,4,5-triphosphate-dependent activation of protein kinase B. *Science* **279**: 710-714

Strasser, A., Huang, D.C.S., Vaux, D.L. 1997. The role of bcl-2/ced-9 gene family in cancer and general implications of defects in cell death control for tumourigenesis and resistance to chemotherapy. *Biochem. Bioph. Acta* **1333**: F151-F178

Sturzbecher, H., Adison, W., and Jenkins, J.R. 1988. Characterization of mutant p53-hsp72/73 protein-protein complexes by transient expression in monkey COS cells. *Mol. Cell. Biol.* **8**: 3740-3747

Swanson, L.W., Kohler, C. and Bjorkland, A. 1987. The limbic region. I. The septohippocampal system, in *Handbook of Chemical Neuroanatomy, Vol. 5: Integrated Systems of the CNS*, A. Bjorkland, T. Hokfelt and L.W. Swanson, eds., pp. 125-277, Elsevier, New York

Takayama, S., Sato, T., Krajewski, S., Kochel, K., Irie, S., Millan, J.A., and Reed, J.C. 1995. Cloning and functional analysis of BAG-1: a novel Bcl-2 binding protein with anti-cell death activity. *Cell* **80**: 279-284

Takayama, S., Bimston, D.N., Matsuzawa, S., Freeman, B.C., Aime-Sempe, C., Xie, I., Morimoto, R.J. and Reed, J.C. 1997. Bag-1 modulates the chaperone activity of Hsp70/Hsc70. *EMBO J.* **16**: 4887-4896

Takayama, S., Krajewski, S., Krajewska, M., Kitada, S., Zapata, S., Zapata, J.M., Kochel, K., Knee, D., Scudiero, D., Tudor, G., Miller, G.J., Miyashita, T., Yamada, M., and Reed, J.C. 1998. Expression and location of Hsp70/Hsc-binding anti-apoptotic Bag-1 and its variants in normal tissues and tumor cell lines. *Cancer Res.* **5**8: 3116-3131

Takayama, S., Xie, Z. and Reed, J.C. 1999. An evolutionarily conserved family of Hsp70/Hsc70 molecular chaperone regulators. *J. Biol. Chem.* **274**: 781-786

Tan, Y., Rouse. J., Zhang, A., Cariati, S., Cohen, P., Comb, M.J. 1996. FGF and stress regulate CREB and ATF-1 via a pathway involving p38 MAP kinase and MAPKAP kinase-2. *EMBO J.* **15**: 4629-4642

Tan, S.-L., Katze, M.G. 2000. HSV.com : maneuvering the internetworks of viral neuropathogenesis and evasion of the host defense. *Proc. Natl. Acad. Sci. USA* **97**: 5684-5686

Teodoro, J.G. and Branton, P.E.1997. Regulation of apoptosis by viral gene products. *J.Virol.* **7**1: 1739-1746

Terada, H., Tsutsui, J., Sanada, J., Arima, T., Ozawa, M. 1997. Heparin binding protein-44 (HBP-44) receptor-associated protein (RAP) mediates cell-substratum adhesion of mouse NIH/3T3 cells through its binding to low density lipoprotein (LDL) receptor-related protein (LRP). *Mol. Membr. Biol.* **14**: 81-86

Tewari, M., Quan, L.T., O'Rourke, K., Desnoyers, S., Zeng, Z., Beidler, D.R., Poirier, G.G., Salvesen, G.S., and Dixit, V.M. 1995. Yama/CPP32beta, a mammalian homolog of CED-3, is a CrmA-inhibitable protease that cleaves the death substrate poly(ADP-ribose) polymerase. *Cell* **81**: 801-809

Theodorakis, P., D'Sa-Eipper, C., Subramanian, T., and Chinnadurai, G. 1996. Unmasking of a proliferation-restraining activity of the anti-apoptosis protein EBV BHRF1. *Oncogene* **12**: 1707-1713

Thomas, K.L., Laroche, S., Errington, M.L., Bliss, T.V., Hunt, S.P. 1994. Spatial and temporal changes in signal transduction pathways during LTP. *Neuron* **13**: 737-745

Thome, M., Schneider, P., Hofmann, K., Fickenscher, H., Meini, E., Neipel, F., Mattman, C., Burns, K., Bodmer, J-L., Schroter, M., Scaffidi, C., Krammer, P.H., Peter, M.E. and Tschopp, J. 1997. Viral FLICE-inhibitory proteins (FLIPs) prevent apoptosis induced by death receptors. *Nature* **386**: 517-521

Thomson, R.L., Sawtell, N.M., Wechsler, S.L., Perng, G.-C., Jones, C., Ghiasi, H., Nesburn, A.B., and Clinton. J. 2000. HSV latency associated transcript and neuronal apoptosis. *Science* **289**: 1651

Tian, Q., Taupin, J.L., Eelledge, S., Robertson, M., Anderson, P. 1995. Fas-activated serine/threonine kinase (FAST) phosphorylates TIA-1 during Fas-mediated apoptosis. *J. Exp. Med.* **182**: 865-874

Tomlinson, A.H. and Esiri, M.M. 1983. Herpes simplex encephalitis. *J. Neurol. Sci.* **60**: 473-484

Tong, L., Toliver-Kinsky, T., Taglialatela, G., Werrbach-Perez, K., Wood, T., and Perez-Polo, J.R. 1998. Signal transduction in neuronal death. *J. Neurochem.* **7**1: 447-459

Tsujimoto, Y. 1998. Prevention of neuronal cell death by Bcl-2, in: *Apoptosis: mechanisms and role in disease.* Kumar S., ed., Springer-Verlag Berlin Heidelberg, pp. 137-155

Uney, J.B., Kew, J.N., Staley, K., Tyers, P. and Sofroniew, M.V. 1993. Transfection-mediated expression of human Hsp70i protects rat dorsal root ganglian neurones and glia from severe heat stress. *FEBS Lett.* **334**: 313-317

Unsicker, K., Grothe, G., Ludecke, G., Otto, D., Westerman, R. 1993. Fibroblast growth factors: their roles in the central and peripheral nervous system, in *Neurotrophic factors*, Loughlin S., Fallon J., eds., Boston, Acad, Press, pp. 313-338

Upton, C., Macen, J.L., Wishart, D.S., McFadden, G. 1990. Myxoma virus and malignant rabbit fibroma virus encode a serpin-like protein important for virus virulence. *Virology* **179**: 618-631

van Dam, H., Duyndam, M., Rottier, R., Bosch, A., de Vries-Smits, L., Herrlich, P., Zantema, A., Angel, P., van der Eb, A.J. 1993. Heterodimer formation of cJun and ATF-2 is responsible for induction of c-jun by the 243 amino acid adenovirus E1A protein. *EMBO J.* **12**: 479-487

van Dam, H., Wilhelm, D., Herr, I., Steffen, A., Herrlich, P. and Angel, P. 1995. ATF-2 is preferentially activated by stress-activated protein kinases to mediate c-jun induction in response to genotoxic stress. *EMBO J.* **14**: 1798-1811

van Hemert, M.J., Steensma, H.Y., and van Heusden, G.P.H. 2001. 14-3-3 proteins: key regulators of cell division, signaling and apoptosis. *BioEssays* **23**: 936-946

Varmeh-Ziaie, S., Okan, I., Wang, Y., Magnuson, K.P., Warthoe, P., Strauss, M., Widman, K.G. 1997. Wig-1, a new p53-induced gene encoding a zinc finger protein. *Oncogene* **15**: 2699-2704

Villalba, M., Bockaert, J., Journot, L. 1997. Concomitant induction of apoptosis and necrosis in cerebellar granule cells following serum and potassium withdrawal. *NeuroReport* **8**: 981-985

Vlahos, C.J., Matter, K.Y. Hui, K.Y., Brown, R.F. 1994. A specific inhibitor of phophatidylinositol 3-kinase, 2-(4-morpholinyl)-8-phenyl-4H-1-benzopyran-4-one (LY294002). *J. Biol. Chem.* **269**: 5241-5248

Vojtek, A.B. and Der C.J. 1998. Increasing complexity of the ras signaling pathway. *J.Biol.Chem.* **273**: 19925-19928

Voll, R.E., Herrmann, M., Roth, E.A., Stach, C. and Kalden, J.R. 1997. Immunosupressive effects of apoptotic cells.*Nature* **390**: 350-351

Wagner, A.J., Kokotis, J.M., Hay, N. 1994. C-Myc-mediated apoptosis requires wild-type p53 in a manner independent of cell cycle arrest and the ability of cells to induce p21[waf1/cip1] genes. *Genes Dev.* **8**: 2817-2830

Wagstaff, M.J., Collaco-Moraes, Y., Aspen, B.S., Coffin, R.S., Harrisson, M.J., Latchman, D.S., de Belleroche, J.S. 1996. Focal cerebral ischemia increases the levels of several classes of heat shock proteins and their corresponding mRNAs. *Brain Res. Mol. Brain Res.* **42**: 236-244

Wagstaff, M.J.D., Collaco-Morales, Y., Smith, J., de Belleroche, J.S., Coffin, R.S. and Latchman, D.S. 1999. Protection of neuronal cells from apoptosis by Hsp27 delivered with a herpes simplex virus-based vector. *J. Biol. Chem.* **274**: 5061-5069

Walton, M., Lawlor, P., Sirimanne, E., Williams, C., Gluckman, P. and Dragunov, M. 1997. Loss of Ref-1 protein expression precedes DNA fragmentation in apoptotic neurons. *Brain Res. Mol. Brain Res.* **44**: 167-170

Walton, M., Woodgate, A.-M., Sirimanne, E., Gluckman, P., Dragunow, M. 1998. ATF-2 phosphorylation in apoptotic neuronal death. *Mol. Brain Res.* **63**: 198-204

Wan, X., Duncan, M.D., Naas, P., Harmon, J.W. 2001. Synthetic retinoid CD437 induces apoptosis of esophageal squamous HET-1A cells through the caspase-3-dependent pathway. *Anticancer Res.* **4A**: 2657-2663

Wang, H.G., Rapp, U.R., and Reed, J.C. 1996. Bcl-2 targets the protein kinase Raf-1 to mitochondria. *Cell* **87**: 629-638

Wang, S., Rowe, M., Lundgren, E. 1996. Expression of the Epstein Barr virus transforming protein LMP1 causes a rapid and transient stimulation of the Bcl-2 homologue Mcl-1 levels in B-cell lines. *Cancer Res.* **56**: 4610-4613

Wang, K., Yin, X.-M., Chao, D.T., Milliman, C.L., and Korsmeyer, S.J. 1996. BID: a novel BH3 domain-only death agonist. *Genes Dev.* **10**: 2859-2869

Wang, H.-G., Pathan, N., Ethel, I.M., Krajewski, S., Yamaguchi, Y., Shibasaki, F., McKeon, F., Bobo, T., Franke, T.F., Reed, J.C. 1999. Ca2+-induced apoptosis through calcineurin dephosphorylation of BAD. *Science* **284**: 339-343

Wang, H.-Y., Lee, D.H.S., D'Andrea, M.R., Peterson, P.A., Shank, R.P., and Reitz, A.B. 2000. β-amyloid 1-42 binds to α7 nicotinic acetylcholine receptor with high affinity. *J. Biol. Chem.* **275**: 5626-5632

Wang, X., Zhang, G.-r., Sun, M., and Geller, A.I. 2001. General strategy for constructing large HSV-1 plasmid vectors that co-express multiple genes. *Biotechniques* **31**: 204-212

Watson, A. and Lowenstein, P. 1998. Therapeutic manipulation of apoptosis in cancer and neurological disease, in *Apoptosis genes*, J.W. Wilson, C. Booth and C.S. Potten, eds., Kluwer Acad. Publishers, pp 281-303

Webster, M.A., Hutchinson, J.N., Rauh, M.J., Muthuswamy, S.K., Anton, M., Tortorice , C.G., Cardiff, R.D., Graham, F.L., Hassel, J.A., Muller, W.J. 1998. Requirement for both Shc and phosphatidylinositol 3' kinase signaling pathways in polyoma middle T-mediated mammary tumorigenesis. *Mol. Cell Biol.* **18**: 2344-2359

Whitley, R.J., Kimberlin, D.W. and Roizman, B. 1998. Herpes simplex viruses. *Clin. Inf. Dis.* **26**: 541-555

Whitley, R.J. and Kimberlin, D.W. 1999. Viral encephalitis. *Pediatric Rev.* **20**: 192-198

Widmann, C., Gibson, S., and Johnson, G.L. 1998. Caspase-dependent cleavage of signaling proteins during apoptosis. *J. Biol. Chem.* **273**: 7141-7147

Wilcox, C.L. and Johnson, E.M. Jr. 1987. Nerve growth factor deprivation results in the reactivation of latent herpessimplex virus in vitro. *J. Virol* . **61**: 2311-2315

Williams, L.R., Varon, S., Peterson, G.M., Wictorin, K., Fischer, W., Bjorklund, A., Gage, F.H. 1986. Continuous infusion of nerve growth factor prevents basal forebrain neuronal death after fimbria fornix transection. *Proc. Natl. Acad. Sci. USA* **83**: 9231-9235

Wilson, T. and Treisman, R. 1988. Fos C-terminal mutations block downregulation of c-fos transcription following serum stimulation. *EMBO J.* **7**: 4193-4202

Wyllie, A.H., Morris, R.G., Smith, A.L. and Dunlop, D. 1984. Chromatin cleavage in apoptosis: association with chromatin morphology and dependence on macromolecule synthesis. *J. Pathol.* **142**: 67-77

Wu, G.-Y., Deisseroth, K. and Tsien, R.W. 2001. Activity-dependent CREB phosphorylation: convergence of a fast, sensitive calmodulin kinase pathway and a slow, less sensitive mitogen-activated protein kinase pathway. *Proc. Natl. Acad. Sci. USA* **98**: 2808-2813

Xia, Z., Dickens M., Raingeaud, J., Davis, R.J., Greenberg, M.E. 1995. Opposing effects of ERK and JNK-p38MAP kinases on apoptosis. *Science* **270**: 1326 - 1331

Xia, K., Knipe, D.M., and DeLuca, N.A. 1996. Role of protein kinase A and the serine-rich region of herpes simplex virus type 1 ICP4 in viral replication. *J. Virol.* **70**: 1050-1060

Xing, J., Ginty, D.D., Greenberg, M.E. 1996. Coupling of the Ras-MAPK pathway to gene activation by RSK2, a growth factor-regulated CREB kinase. *Science* **273**: 959-963

Xue, D., Horvitz, H.R. 1995. Inhibition of the Caenorhabditis elegans-cell death protease CED-3 by a CED-3 cleavage site in baculovirus p35 protein. *Nature* **377**: 248-251

Yamada, T., Yoshiyama, Y. and Kawaguchi, N. 1997. Expression of activating transcription factor-2 (ATF-2), one of the cyclic AMP response element (CRE) binding proteins, in Alzheimer disease and non-neurological brain tissues. *Brain Res.* **749**: 329-334

Yamaguchi, A., Tamatani, M., Matsuzaki, H., Namikawa, K., Kiyama, H., Vitek, M.P., Mitsuda, N., and Tohyama, M. 2001. Akt activation protects hippocampal neurons from apoptosis by inhibiting transcriptional activity of p53. *J. Biol. Chem.* **276**: 5256-5264

Yang, K., Mu, X.Z., Xue, J.J., Whitson, J., Salminen, A., Dixon, C.E., Lui, P.K. and Hayes, R.L. 1994. Increased expression of c-fos mRNA and AP-1 transcription factors after cortical impact injury in rats. *Brain Res.* **664**: 141-147

Yang, E., Zha, J., Jockel, J., Boise, L.H., Thomson, C.B., and Korsmeyer, S.J. 1995. Bad, a heterodimeric partner for Bcl-xL and Bcl-2, displaces Bax and promotes cell death. *Cell* **80**: 285-291

Yano, S., Tokumitsu, H., and Soderling, T.R. 1998. Calcium promotes cell survival through CaM-K kinase activation of the protein-kinase-B pathway. *Nature* **396**: 584-587

Yao, F. and Courtney, R.J. 1991. Association of a major transcriptional regulatory protein, ICP4, of herpes simplex virus type 1 with the plasma membrane of virus-infected cells. *J. Virol.* **65**: 1516-1524

Yao, R., and Cooper, G. M. 1995. Regulation of the Ras signaling pathway by GTPase-activating protein in PC12 cells. *Oncogene* **11**: 1607-1614

Yeh, H.Y., Itie, A., Elia, A.J., Ng, M., Shu, H.B., Wakeham, A., Mirtsos, C., Suzuki, N., Bonnard, M., Goeddel, D.V., Mak, T.W. 2001. Requirement for Casper (c-FLIP) in regulation of death receptor-induced apoptosis and embryonic development. *Immunity* **12**: 633-642

Yew, P.R., Liu, X. and Berk, A.J. 1994. Adenovirus E1B oncoprotein tethers a transcriptional repression domain to p53. *Genes Dev.* **8**: 190-202

Yip-Schneider, M.T., Miao, W., Lin, A., Barnard, D.S., Tzivio, G., Marshall, M.S. 2000. Regulation of the Raf-1 kinase domain by phosphorylation and 14-3-3 association. *Biochem. J.* **351**: 151-159

York, R.D., Yao, H., Dillon, T., Ellig, C.L., Eckert, S.P., McCleskey, E.W., Stork, P.J. 1998. Rap1 mediates sustained MAP kinase activation induced by nerve growth factor. *Nature* **392**: 622-626

Yuan, J. and Yankner, B.A. 2000. Apoptosis in the nervous system. *Nature* **407**: 802-809

Yun, H.Y., Gonzales-Zulueta, M., Dawson, V.L., Dawson, T.M. 1998. Nitric oxide mediates N-methyl-D-aspartate receptor-induced activation of p21ras. *Proc. Natl. Acad. Sci. USA* **95**: 5773-5778

Zachos, G., Clement, B., and Conner, J. 1999. Herpes Simplex Virus Type 1 infection stimulates p38/c-jun N-terminal mitogen-activated protein kinase pathways and activates transcription factor AP-1. *J. Biol. Chem.* **274**: 5097-5103

Zachos, G., Koffa, M., Preston, C.M., Clements, J.B., and Conner, J. 2001. Herpes simplex virus type 1 blocks apoptotic host cell defense mechanisms that target Bcl-2 and manipulates activation of p38 mitogen-activated protein kinase to improve viral replication. *J. Virol.* **75**: 2710-2728

Zha, J., Harada, H., Yang, E., Jockel, J., and Korsmeyer, S.J. 1996. Serine phosphorylation of death agonist BAD in response to survival factor results in binding to 14-3-3 not bcl-xL. *Cell* **87**: 619-628

END OF TEXT

www.ingramcontent.com/pod-product-compliance
Lightning Source LLC
Chambersburg PA
CBHW081106170526
45165CB00008B/2340